NATURAL ENVIRONMENT RESEARCH COUNCIL
INSTITUTE OF GEOLOGICAL SCIENCES

British Regional Geology

South-West England

(FOURTH EDITION)

By E. A. Edmonds, M.Sc.
M. C. McKeown, B.Sc.
M. Williams, B.Sc., Ph.D.

Based on previous editions by
H. Dewey

LONDON   HER MAJESTY'S STATIONERY OFFICE   1975

*The Institute of Geological Sciences
was formed by the incorporation of the
Geological Survey of Great Britain
and the Museum of Practical Geology
with Overseas Geological Surveys
and is a constituent body of the
Natural Environment Research Council*

© *Crown copyright 1975*

First published 1935
Fourth edition 1975

1 SBN 0 11 880713 7

Foreword to the Fourth Edition

The third edition of this handbook, published in 1969, retained the general arrangement adopted by the late Mr. H. Dewey in the first edition (1935). Chapters 2 and 3 were written by Mr. M. C. McKeown, Chapter 6 and the metalliferous mining section of Chapter 10 by the late Dr. M. Williams, and the rest by Mr. E. A. Edmonds. The authors gratefully acknowledged the advice and assistance given by many colleagues, including Mr. J. E. Wright (Carboniferous rocks), Dr. J. R. Hawkes (igneous rocks and written notes on contact metamorphism), Mr. C. J. Wood (written notes on Cretaceous rocks), Mr. K. E. Beer (mining) and Mr. G. Bisson (District Geologist, South-West England), as well as Professor W. R. Dearman and the contributors to the special commemorative volume [1966] of the Royal Geological Society of Cornwall.

The succeeding years have seen notable advances in our knowledge of the geology of south-west England. Most of the great expanse of Carboniferous rocks has been mapped and a regional classification of these strata has been evolved. For the fourth edition Mr. Edmonds has re-written much of the Carboniferous and Pleistocene accounts, added a chapter on structure and made smaller amendments throughout the book. The work has again been edited by Mr. Bisson.

As world resources of minerals that form the physical basis of civilisation are increasingly used up and as oil companies turn their attention to the seas west of Britain, so attention is increasingly focused on the south-western peninsula. The successful blending of people, industry and amenity, while leaving the striking scenery and landscape unimpaired, depends upon a thorough knowledge of the geology and structure of the region, of which this book provides a summary.

KINGSLEY DUNHAM
Director

Institute of Geological Sciences,
Exhibition Road,
London SW7 2DE
10th June 1974

An EXHIBIT illustrating the geology and scenery of the region described in this handbook is set out in the Geological Museum, Institute of Geological Sciences, Exhibition Road, South Kensington, London SW7 2DE.

Contents

Illustrations

Figures in Text

Plates

[1]Numbers preceded by A or MN refer to photographs and negatives in the Geological Survey collections.

I. Introduction

Position, Climate and Scenery

The area of south-west England described in this handbook (Fig. 1) embraces Cornwall, Devon, part of Somerset and the Scilly Isles; it is bounded by the Atlantic, the Bristol Channel and the English Channel, and to the east by an arbitrary line from the Parrett estuary to Charmouth.

Nowhere in the peninsula is the sea more than 25 miles distant and the maritime climate is characterized by milder winters and cooler summers than obtain over much of Britain. However, the strong, predominantly westerly winds bring much rain, about 40 inches a year in low-lying inland areas and more than 80 inches on high Dartmoor, and the popular phrase 'English Riviera' can be applied with some justice only to south Cornwall and the Scillies.

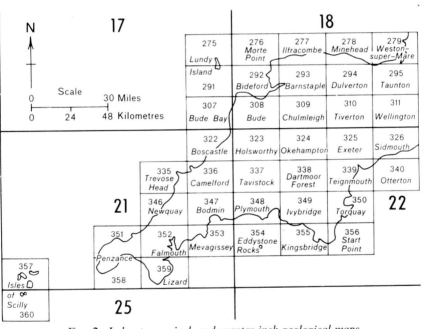

FIG. 2. *Index to one-inch and quarter-inch geological maps*

Rugged cliffs of the north coast, which are among the most magnificent in England and Wales, contrast with gentler hills rolling down to the English Channel. Headlands are formed of the harder sediments and volcanic rocks and also commonly of dykes or larger intrusive masses. The coast, largely of old hard rocks and much indented by drowned river mouths, is at a young stage of shoreline evolution.

Inland, sixteen Cornish hills rise to over 1000 ft, the highest being Brown Willy (1375) on Bodmin Moor. Yes Tor (2027) and High Willhays (2038), on Dartmoor, are the highest points in southern England, and in north Devon and nearby Somerset Exmoor (Dunkery Beacon 1707), the Brendon Hills (Lype Hill 1390) and the Quantocks (Will's Neck 1261) form high ground of Devonian rocks. Hard sandstones of Devonian and Carboniferous age have given rise to long ridges separated by shale valleys, and wider tracts of shale country are marked by bogs, marshes and heathy moors. Of the younger rocks to the east, the New Red Sandstone of south Devon and west Somerset yields some of the richest agricultural land in the region.

Most rivers flow south to the English Channel. The main Cornish exception is the Camel. In Devon the Taw and the Torridge converge on Bideford Bay, and the small Lynn is perhaps worth noting because of the Lynmouth flood disaster of 1952. South-flowing rivers in Cornwall include the Fal and the Fowey, and the Tamar forms the county boundary almost from coast to coast. Farther east, between the Tavy and the Teign, several streams radiate southwards from Dartmoor. The Exe, Devon's main river, rises in Somerset and pursues a fairly constant southerly course to Exmouth. Still farther east the Otter and the Axe drain the area where Somerset, Devon and Dorset meet.

In terms of resources there is no water supply problem in south-west England. Most supplies are taken from surface waters and several reservoirs have been completed in recent years. Thus the Meavy has been impounded at Burrator to supply Plymouth, and a reservoir at Huntingdon Warren, on the Avon, was completed in 1960. Torquay draws water from a reservoir at Fernworthy on Dartmoor where the South Teign has been dammed. Bridgwater is supplied from Hawkridge Reservoir, which lies on the east side of the Quantocks and first overflowed in 1962. Small hydro-electric plants have been built at Marytavy, Morwellham and Chagford, on the Tavy, the Tamar and the Teign.

Man and Industry

Stone Age man, with his simple tools, lived mainly near water or on unforested uplands. In Neolithic times he built stone burial chambers, or dolmens, one of the best preserved of which is Spinster's Rock, Drewsteignton, and commonly heaped earth over them to form barrows. Methods of stock and arable farming were probably introduced by Mediterranean immigrants during the fourth millenium B.C.

People from the eastern Mediterranean brought metals, metal-working technology and a knowledge of prospecting to inaugurate the Bronze Age, when some export of tin occurred. Use of wood as fuel for smelting began the deforestation of the south-west, thus making ever-increasing areas of land available for agriculture. The hut circles, avenues, pounds, kistvaens and menhirs, characteristic of moorland, probably date from Bronze Age times. People of the hut settlements were no doubt in part farmers, but a striking coincidence of such communities with old workings suggests that many were pioneer tin-streamers.

Celtic immigrants began to arrive in the Bronze Age and gradually established Iron Age culture. Their iron tools and consequent improved agricultural techniques led to the establishment of more permanent settlements. Hill-fort earthworks were thrown up on many ridges and hilltops.

The Romans established a garrison community at Exeter and built a few roads and small stations farther west. Subsequently the Anglo-Saxons settled as far west as Devon and Viking invaders founded communities along a few narrow coastal strips, but throughout all this time Celtic peoples remained the dominant race in Cornwall. With the unification of England under Norman rule began the disintegration of such racial distinctions.

The concentration of people around mines is reflected in the fact that the Camborne–Redruth district is still the most densely populated large area in Cornwall. Penzance and Truro both owe their early expansion largely to nearby mining, although Truro, centrally placed and a busy port since the Middle Ages, has long been an important market town. With the decline of tin mining, St. Austell developed as the centre of the china clay industry. Bodmin superseded Launceston as county town of Cornwall in 1838, but it is not a good rail centre and lies east of the main concentration of population; consequently Truro has become the main administrative centre. The fishing industries of St. Ives and Newquay have given way respectively to art and the tourist trade.

No deep-water harbours are present on the north coast, but in the south Falmouth affords almost 10 square miles of sheltered anchorage, and Charlestown, Par and Fowey have long handled the export of china clay from the St. Austell district. Plymouth's naval importance dates back to the foundation of Devonport dockyard by William of Orange, but the town developed surprisingly late as an important port, more or less with the decline of river traffic to Exeter, which city, however, remains the major centre of commerce and administration in the west country. Minor ports include Exmouth and Dartmouth. Several small coastal or estuary towns were once busy ports, for example Penzance, St. Ives, Bideford, Barnstaple and Watchet, but all are now largely dependent on the holiday trade. Bridgwater, a small port 10 miles from the mouth of the Parrett, carries on a small coastal and overseas trade and is one of the few west country harbours whose importing of coal survived the opening of the Severn Tunnel. The unique bath bricks, of silt, sand and clay from the river, are still manufactured here. Colliers from Swansea still call at Hayle, once the main ore-exporting port, and shipbuilding is carried on at Appledore. Inland, Taunton is a rail junction, a market town, and the county town of Somerset.

Railways arrived late in south-west England and have never provided an adequate network. Some individual enterprises added local lines, as at Moretonhampstead, Lynton and the Haytor granite tramway, but none survived and now branch lines are becoming fewer. Roads are being improved and new ones built; in 1961 a great new suspension bridge over the Tamar, intended to replace the Saltash ferry, was opened. Of several inland waterways the Exeter Canal, the Taunton–Bridgwater canal and part of the Tiverton–Taunton canal are still open.

In the main, south-west England is devoted to dairy and stock farming, sheep, pigs and poultry, and holdings are commonly small. Large areas of arable land are generally confined to the post-Carboniferous rocks in the east. Some good arable farmland exists elsewhere, especially in the south, away from the high ground and moors, but much of the interior is ill drained and unproductive. Sheltered valleys in the south afford rich alluvial soils for the growing of spring flowers and soft fruit; horticulturists in the Mount's Bay area raise three crops a year, and Cornwall and the Scilly Isles 'export' 250 million cut flowers a year. Lime-rich beach-sand is collected for use as a fertiliser and the Bude Canal, the longest in Cornwall, was built to carry such sand inland. The seaweed now spread on coastal land as a source of potash was, between the 17th and early 19th centuries, burned in kelp kilns to produce low-grade soda. Such kilns were particularly common on the Scillies and the soda was sent to industrial centres as far away as London and the Midlands for use in making glass, soap and bleaching powder.

On the high moorland the soil is thin and acid and affords grazing to Scottish hill sheep and, in the summer, Galloway cattle. South Devon and Somerset are famous for cider apples. In 1919 the newly established Forestry Commission planted its first trees, near Eggesford in the valley of the Taw, and since then afforestation has proceeded as fast as money, land and permission could be obtained. Original oak forest now remains as small scattered woods, commonly along valleys, and two of the three stands of stunted but ancient oaks on Dartmoor, Black-a-Tor Copse and Wistman's Wood, have been designated Forest Nature Reserves. Land acquired for afforestation commonly has a soil so thin and poor that only after one, or even two crops of conifers, will mixed planting be possible and commercial plantings are almost entirely of spruce, fir and pine. Some beech and oak are being planted with conifers in the Quantocks. The Scots Pine is described in the 'Flora of Devon' as perhaps originally indigenous but despite this the association of conifer and heather, acknowledged as pleasing on the heath-lands of Surrey and Hampshire, has encountered some opposition in the south-west. Perhaps this will diminish now that thought is being given to fitting forests into the landscape rather than imposing arbitrary rectangular plantations, and to encouraging visitors to ramble along the rides. The Fernworthy Reservoir is most attractively set amid conifer-clad slopes. Throughout the region forestry is affording employment on much marginal land, and elsewhere it can be carried on in face of the urban drift of population since it requires fewer workers per square mile than does most farming.

Fishing as well as mining was once a major industry in south-west England, particularly in Cornwall whence pilchards were sent to the West Indies and the Mediterranean. Some mackerel is still fished, but most angling is now done for sport.

The tin streamers were exhausting the reserves of alluvial tin by the 15th century, but underground mining was then already common along the ten-mile-wide mineralized belt which stretches from Land's End to Tavistock. Large copper reserves were known in the 18th century and by the mid-19th century some 400 mines were active, but cheaper ore from overseas gradually killed the industry. Enforced but temporary revival took place during the two world wars and between 1939 and 1945 five mines were operating. Recently intensified prospecting has increased the number of active mines from two to four. Iron ore was mined near Lostwithiel, Perranporth and South Molton and the West Somerset Mineral Railway was opened in 1857 to carry ore from quarries on Brendon Hill to Watchet. In recent years hematite has been worked at Hennock, baryte at Bridford and wolfram at Castle-an-Dinas. The working of lead, largely for associated silver, was of some importance in Tudor times, and a little gold was mined at Ladock.

The huge slate quarry at Delabole has been active since the 16th century, but roofing slate is now so expensive that much of the output goes to form inert granular additives for incorporation in such things as fertilisers, plastics, roofing felt and concrete blocks. Granite and limestone have similarly been superseded as building stones although both are still used locally. Burning of limestone for lime, formerly carried on at numerous small quarries, is no longer widespread, but some lime is still produced. A variety of rocks is worked for roadstone, aggregate and ballast, as at Meldon Quarry, which is one of the largest in the country. Sand and gravel are also dug, and of the many clays used in the past and present for everything from cob walls to fine china may be mentioned those of Bovey Tracey and Petrockstow. These ball clays or pottery clays, upon which was founded the once famous Bovey pottery, are now an important export. Lignite from these beds has been worked for fuel. China clay, separated as a white kaolin mud by lavation from the quartz and mica of decomposed granite, is worked in the St. Austell area, on the edges of Bodmin Moor and at Lee Moor on the south-western slopes of Dartmoor and is a major export. The huge white quartz-rich tip-heaps are a great disfigurement of the landscape.

The necessary combination of suitable foundation, abundant available water for cooling, and sparse population has permitted the building of a nuclear power station near Hinkley Point, on the fens bordering Bridgwater Bay. It is the first such station in the world to produce $\frac{1}{2}$ million kilowatts for civil use, and its construction has provided some employment and brought about many road improvements and the rebuilding of Combwich harbour. A second power station, using an Advanced Gas Cooled Reactor, is being built in the same area and will generate $1\frac{1}{4}$ million kilowatts.

The tourist trade is second in importance only to agriculture as an industry in south-west England. Most holidaymakers head for the coast whose indentations provide excellent waters for yachting, as in Bridgwater Bay, Bideford Bay, the Camel estuary, Mount's Bay and along extensive stretches of the south coast. Sea fishing is becoming increasingly popular and provides employment for boats and fishermen of the decaying fishing industry. Inland, all the major rivers yield brown trout, sea trout and salmon and most of the popular fresh-water fishes have been recorded. Boating and fishing are practised on Hawkridge Reservoir and it is surprising that more use is not made of reservoirs as sports centres.

The several abandoned canals of south-west England afford some fishing, and some short lengths may still be navigable.

Limestone areas between Ashburton and Tor Bay, and near Plymouth, afford facilities for caving, while Kent's Cavern near Torquay has yielded ancient bones, both human and animal. On the east side of the Quantocks, just south of Aisholt, Holwell Cavern was once noted for its stalactites and stalagmites but many were destroyed by visitors and the cave is no longer open to the public.

Naturalists and ramblers have been attracted to the south-west for many years and National Parks have been established on Dartmoor and Exmoor. The last of Dartmoor's resident red deer were killed towards the end of the 18th century but they are plentiful on Exmoor, less so on the Quantocks; fallow deer and a few roe deer are widespread in central Devon and the Japanese sika deer has been recorded in north Devon and on neighbouring Exmoor. More attractive to visitors, because relatively tame, are the ponies which roam Dartmoor, Exmoor, the Brendon Hills and the Quantocks.

Bodmin Moor, the Quantocks and long stretches of coast have been declared Areas of Outstanding Natural Beauty. Coastal National Nature Reserves have been declared around the Parrett estuary (Bridgwater Bay), at Braunton Burrows (near Barnstaple) and between Axmouth and Lyme Regis, and inland ones at Yarner Wood on the east side of Dartmoor and in the Bovey valley. Wildfowl refuges have been established at Chapel Wood near Barnstaple, Walmsley on the River Camel, between the Exe estuary and Dawlish Warren, and on Annet in the Scilly Isles.

Geological History

With the exception of parts of the basic igneous complex of the Lizard, which may be of Pre-Cambrian age, the oldest rocks of the peninsula are relatively small exotic masses of Ordovician quartzite and Silurian limestone lying within Devonian beds at Meneage and Nare Head. In Ordovician times the area lay beneath a fairly shallow warm sea which deepened northwards towards an east–west trough stretching through what is now central Wales. This shallow sea extended during the Silurian period to cover a land mass stretching from the Bristol Channel across eastern England.

The late Silurian–early Devonian Caledonian orogeny raised land to the north of the present peninsula to produce a semi-desert which drained southwards through swamps and deltas to the geosyncline of the Devonian Sea. In deltas and sea accumulated the deposits which formed the Lower Devonian, of sandstones, shales, conglomerates and some calcareous beds. Succeeding Middle and Upper Devonian slates and mudstones represent the finer deposits consequent upon the northern land being worn down to a low rolling plain; local clear shallow water permitted the building of coral banks, as evidenced by the limestones between Plymouth and Torquay and near Aisholt in the Quantocks.

Volcanic activity of the time produced widespread submarine lava-flows which commonly present pillow forms. They are associated with tuffs and extend from the north coast of Cornwall, where they form the towering cliffs of Pentire, eastwards for about eight miles; a second strip extends from the south of Bodmin Moor past Liskeard and Plymouth to Ashprington and Tor Bay. There seems to be a close relationship between these lavas and the limestones, for where one series dies out the other appears and it may be that volcanic vapours locally prevented the growth of coral reefs.

During Carboniferous times silt and mud, commonly limy, formed the rocks of the Lower Carboniferous, and great outbursts of vulcanicity are now represented by lavas, tuffs, ashes and agglomerates. Thick beds of radiolarian chert, generally forming long ridges, are associated with lenticular black limestone. As the sea grew shallower, and even locally swampy, river-borne sand and mud with some carbonaceous matter accumulated in Upper Carboniferous times and the resultant sandstones and shales now occupy the greater part of central Devon, in the heart of the huge synclinorium which is the key to the structure of the area (Fig. 3). Local earth movements caused changes in level and consequent changes in sedimentation; peaty layers which developed on mud flats just above sea level were quickly covered by silt, sand and mud following slight submergence. The waters of the marine basin were frequently disturbed by turbidity currents, particularly in early Upper Carboniferous times.

Towards the close of the Carboniferous Period the immensely thick sediments of the Devonian–Carboniferous sea were folded along east and west lines by movements of the Armorican orogeny. Forces were sufficient to produce overfolding, faulting and thrusting and the development of slaty cleavages. Intrusion of granite and consequent metamorphism of the country rock occurred during the closing stages of these movements, and a single large batholith is now exposed as six cupolas, five on the mainland and one forming the Scilly Isles; a seventh (Haig Fras) crops out under water. The cupolas, and many of the mineral veins associated with the granite, appear to follow the N.E.–S.W. trend typical of Caledonian fold axes; this is particularly evident if allowance is made for Tertiary dextral wrench-faulting.

In the immediate post-Armorican scene south-west England was part of an extensive land surface. The final phase of continental drift had not begun and the equator lay perhaps only two or three hundred miles away. The climate was that of a desert in which wet-season torrents washed rock debris into basins and huge screes accumulated at the foot of hill slopes. Thus were formed the Permian breccias and breccio-conglomerates, while marls and sandstones were laid down in fresh water and volcanic activity broke out on a minor scale. Non-marine breccias, pebble-beds, sandstones and marls of Triassic age were laid down during continuing arid conditions.

Most of the region may have been land during the Jurassic period, forming part of the great North Atlantic continent of Laurasia which was then drifting north and beginning to split into two. A warm shallow sea spread over much of the rest of what is now southern England. Uplift in Lower Cretaceous times caused shallowing of this sea and the development over southern England of delta swamps into which rivers flowed from the west. Later the sea advanced again as far west as the eastern edge of Dartmoor; glauconitic sands deposited in the shallower water formed the greensands, and mud in less disturbed or deeper water formed the Gault. Giant reptiles roamed on the land and in the shallows and swam in the sea. The Lower Cretaceous landscape was gradually eroded to one of low relief, and when further subsidence allowed the spread of the great Chalk Sea few rivers were sufficiently active to deposit noticeable quantities of terrigenous material on the sea bed. Instead there accumulated vast quantities of calcareous ooze which consolidated into chalk. Slow break-up of the vast land mass of Laurasia was probably largely complete by mid-Cretaceous times.

Early Tertiary uplift caused retreat of the water, and the Lundy Granite was intruded in Eocene times. Land including the area of the present western England and Ireland drained southwards and eastwards through large deltas to the sea. But in late Oligocene times further subsidence occurred, and certain deposits of clay, sand and gravel with lignite, as at Bovey Tracey and Petrockstow, were formed in rivers and lakes at this time.

The Alpine orogeny, which reached a peak during the Miocene Period, raised the whole of the British Isles above sea level and caused the shearing of south-west England by a number of largely dextral wrench faults.

Subsequently the Pliocene North Sea spread over parts of southern England, lapping against the southernmost parts of the present land of the south-west and perhaps depositing the clays and sands of the St. Erth Beds, although the age of these deposits is in doubt. Warm at first, the waters became gradually colder with the approach of the Ice Age.

Marine erosion in Tertiary times left several peneplains; indeed a multiplicity of such platforms has been reported by various writers. Of the upland flats at 1000 ft, 750 ft and 430 ft O.D. the first might possibly pre-date the Alpine movements and the second is probably Miocene. The lowest platform, a wide, gently shelving plain cut probably in Pliocene times, is one of the characteristics of coastal topography. Rapid emergence of land was much assisted by the enormous quantities of water being increasingly locked up in northern ice.

Another marine platform, on which was deposited the Raised Beach, was cut during an early part of the glacial period, probably before the Pleistocene ice sheets, spreading towards the Bristol Channel area, had reached the northern edge of the present Devon and Somerset. Subsequently the whole of the peninsula became locked in a semi-arctic climate. Snow lay permanently on the uplands and small local glaciers crept down from the highest peaks; fragmentation of outcrops provided the debris of present day clitters. Arctic mammals roamed the tundra browsing on the grasses and sedges, mosses and lichens, which survived the intense cold, and perhaps also on dwarf birches and willows.

The previous elevations of the land, indicated by the marine platforms, had led to successive deepenings of the valleys, upon whose sides the remnants of old alluvial deposits remained as terraces. Then with recessions of the ice and the consequent rises in sea level and regrading of rivers these ancient valleys became filled with alluvial silt, clay and peat up to 120 ft thick; forests flourished upon the rich soils of the valley floors. Readvances of the ice, with consequent falls in sea level and rapid downcutting by rivers, further assisted in terrace formation. With the last retreat of the ice great volumes of water were released, flooding the lower reaches of the valleys to form rias, as at the mouths of the Tamar, the Fal and the Camel. The carbonized remains of forests which grew on the alluvial soils of such valleys are often exposed by storms removing the overlying sea-sand. The rise in sea level continued into Recent times, as the ice front retreated farther and farther north, and the final flooding of the estuaries of Cornwall and Devon may have coincided with the breaching and submergence of the land bridge which united Britain and France across the Strait of Dover, perhaps scarcely 10 000 years ago.

Rising sea level fathered the legend of Lyonesse, a fertile land said to have lain between Land's End and the Scillies around a great city on the site of the Seven Stones. These granite rocks, now awash at about three-quarter tide and upon which the great oil tanker Torrey Canyon foundered in the Spring of 1967, may have been an island in the not too distant past. However,

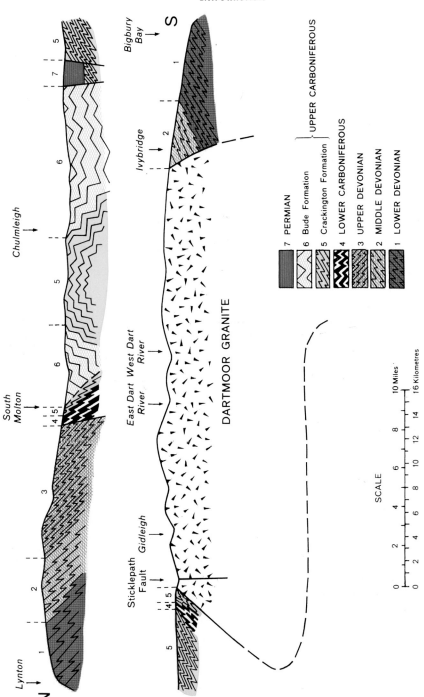

FIG. 3. *Diagrammatic section across Devonshire showing the probable underground shape of the Dartmoor Granite*

Lyonesse was surely Scilly. At low tide it is even now possible to walk between several of the main islands and to see field walls extending out to sea. Thus within the time of 'man the farmer' the Scillies may have comprised a single large island with a few rocky islets—the Siluram Insulam of the Roman writers.

History of Research

Among the early geological observers the name of Borlase ranks first. His 'Natural History of Cornwall' (1758), which is of more than merely historical interest, gives a good general account of the rocks and minerals of the county and his acute powers as an observer are shown by his noting the occurrence of the bed of pebbles below the (Pliocene) deposits at St. Agnes Beacon.

Of more importance to the mining student is Pryce's 'Mineralogia Cornubiensis', published in 1778. In the beginning of the 19th century the 'Transactions of the Royal Geological Society of Cornwall' appeared, and among many annual volumes of interest two in particular stand out as classics from the mining point of view, the first being Henwood's treatise 'On the Metalliferous Deposits of Cornwall and Devon', published as Volume 5 of the Transactions, and the other, the Volume for 1912, by Collins on 'The West of England Mining Region'.

Conybeare was one of the earliest observers to classify the rocks of Cornwall and Devon, in 1814 and 1823.

The official survey of Cornwall and Devon was some of the first work undertaken by the Geological Survey. When De la Beche was appointed the first Director of that institution in 1835, of the eight sheets of Devon and Somerset on which he was engaged four were published, three others completed and the eighth nearly completed.

In 1836 Sedgwick and Murchison commenced researches in Devonshire; they separated the Carboniferous rocks (Culm Measures) from the general mass of deposits which De la Beche had described under the term 'Grauwacke', and (1840) introduced the term Devonian for part of the grauwacke between the Silurian and the Carboniferous. In 1839 De la Beche's classic 'Report on the Geology of Cornwall, Devon and West Somerset' was published and in 1846 ('Essays', *Mem. Geol. Surv.*, vol. i, p. 50) he stated "in Cornwall and Devon there may be equivalents both of the Carboniferous limestone above and of the higher parts of the Silurian beneath also included in this [Devonian] system". In his report he gave credit to Mr. Henry McLauchlan and Mr. Henry Still, Ordnance Surveyors, for their able assistance in delineating the mineral veins, elvans and granite margins and for communicating to the Survey "a mass of important geological facts". Dr. S. R. Pattison of Launceston and the Rev. R. Hennah had helped with the collection of fossils, while the Rev. David Williams and Mr. T. Weaver had indicated some of the main stratigraphical divisions among the older rocks.

By 1840 De la Beche had issued a revised edition of the maps of Cornwall and Devon whereon he adopted divisions of the strata similar to those made by Sedgwick and Murchison as to the order of sequence, applying provisionally to the Culm rocks the term 'Carbonaceous' series and to the Devon and Cornish slates the name 'Grauwacke'.

Sedgwick went to some lengths to explain why it was that he and Murchison had not established the Devonian System in 1836 when they determined the great Culm trough of north Devon to be Carboniferous. His words are: "We sent a good series of the fossils of Petherwin and Barnstaple Groups to London. They were examined and named and every species was called Silurian . . . On re-examining the fossils in 1838 it turned out that all the species of the Barnstaple group had been wrongly named; and that so far from being Silurian, the only doubt respecting them was, whether they might not be called Carboniferous rather than Devonian."

Geological research work was carried on between the years 1855 and 1873 by W. Pengelly and A. Champernowne in south Devon, while H. B. Holl studied the older rocks of the same area. They largely followed the work of Godwin-Austen, who in 1842 combined four of his previous papers into a connected description entitled 'On the geology of the south-east of Devonshire'; this paper may be regarded as the foundation on which all subsequent work in the area was built. The fossil fishes of the Polperro and Looe Beds were studied by C. W. Peach, A. Q. Couch, W. Pengelly and the Rev. David Williams, and as a result of their work the true stratigraphical position of the Dartmouth Slates was determined as Lower Devonian.

In later years much work was done by R. N. Worth, Howard Fox and Upfield Green, while the petrography and chemistry of the igneous rocks and mineral veins received exhaustive treatment by J. A. Phillips.

The official revision of De la Beche's maps was commenced by W. A. E. Ussher in 1870 in the Wellington district. He gradually worked westward and in 1873 had reached Exeter and by 1875 the area around Torquay, where his associates were H. B. Woodward and Clement Reid. Ussher pressed forward through the Bolt and Start country and by Ivybridge and Modbury to Plymouth and St. Austell. He thus surveyed a very large part of south Devon and Cornwall, and by his knowledge of the Devonian of Belgium and western Germany was able correctly to correlate the divisions of the Devonian system of the several areas.

South Cornwall was resurveyed by J. B. Hill, who divided the killas of the Falmouth district into several lithological types and assigned to them an early Palaeozoic age. J. B. Wilkinson and E. E. L. Dixon co-operated with him in western Cornwall together with Clement Reid, while D. A. MacAlister was responsible for the mineral survey.

Petrology of Cornish and Devonian rocks, as noted above, was dealt with by Phillips and also by F. Rutley and S. Allport, but T. G. Bonney was the first to attack the problem of the Lizard rocks by the methods of modern petrography (1877). His work and inferences raised a controversy in which J. H. Collins, H. Fox, C. A. McMahon and J. J. H. Teall took part. A. Somervail and Harford Lowe also contributed papers, Lowe settling the age of the granulitic gneisses. The study of the foliation of the Lizard rocks was advanced by Bonney, Teall and McMahon. Bonney and McMahon ascribed this process to injection foliation in a viscous rock, whereas Teall appealed to dynamic metamorphism. It remained for Sir John S. Flett to explain that the structures could only have been produced by both these agencies acting together. Flett also described the petrography of the mineral veins, the process of kaolinization and allied changes, the granites and their aureoles

Table of formations present in south-west England

Years ago (millions)

QUATERNARY—			
Recent			Beach deposits; Peat; Alluvium
	—0·01—		Boulder clay; Raised beach deposits } River gravels; Head
Pleistocene			
	—3—		
TERTIARY—			
Pliocene?		St. Erth Beds	Clay and sand
	—12—		
Oligocene		Bovey Beds	Clay, sand, gravel and lignite
	—39—		
Eocene			Gravel, sand and clay
	—70—		
MESOZOIC—			
		⌠ Upper Chalk	Nodular chalk with prominent bands of black flints
		Middle Chalk	Nodular chalk passing up into firm white chalk with scattered flints; Beer Freestone locally at base
Cretaceous		Lower Chalk	Soft white chalk; Cenomanian Limestone or sands locally
		Upper Greensand	Sands; Calcareous Grit locally at top
		⌊ Gault	Silt and silty clay
	—135—		
Jurassic		Lower Lias	Clay with limestone
	—180—		
		⌠ Rhaetic	Shale, silt and limestone
Triassic		Keuper Marl	Marls, locally with rock-salt
		Keuper Sandstone	Coarse sandstone
		⌊ Bunter Sandstone	Pebble beds
	—225—		
PALAEOZOIC—			
		⌠ Crediton Conglomerates	⌠ Red Marls
		Knowle Sandstones	Dawlish Breccias
Permian			} Teignmouth Breccia
		Bow Conglomerates	Watcombe Conglomerate
		Cadbury Breccia	⌊ Watcombe Clay
	—280—		
			⌠ Massive sandstones with shales (Bude Formation)
		Upper Carboniferous	Shales with thin turbidite sandstones (Crackington Formation and equivalent Welcombe Formation)
Carboniferous			
		Lower Carboniferous	Shales, sandstones, cherts and limestones
	—350—		
		⌠ Upper Devonian	
Devonian		Middle Devonian	} Slates, mudstones, sandstones and limestones
		⌊ Lower Devonian	Sandstones, shales, conglomerates and some calcareous beds
	—400—		
Silurian		Limestone	} Exotic blocks within
	—440—		Devonian rocks
Ordovician		Quartzite	
	—500—		
IGNEOUS AND HIGH-GRADE METAMORPHIC ROCKS—			
Extrusive			{ Tuff, ash and agglomerate / Lava
			⌠ Granite
			Aplite, pegmatite and porphyry
Intrusive			Lamprophyre
			Dolerite ('greenstone')
			Gabbro
			⌊ Peridotite, serpentine and picrite
Metamorphic			Gneiss, mica-schist and hornblende-schist

and the greenstones and spilites. G. Barrow and H. Dewey contributed to the descriptions of these rocks, and with C. Reid, J. B. Scrivenor and R. L. Sherlock mapped a wide area of north Cornwall and west Devon.

Dr. E. M. Lind Hendriks (1931; 1937) contributed much to our knowledge of the rocks of south Cornwall and later (1959) summarized her views on the structure of south-west England. Gravity and magnetic surveys by Bott and others (1958) led to the interpretation of deep-seated structure, and detailed local structural studies were carried out by Wilson (1946; 1951), Blyth (1957; 1962) and Dearman (1959; 1964). The search for radioactive minerals by the Atomic Energy Division of the Geological Survey reached a peak between 1957 and 1960, and a combined electromagnetic, radiometric and magnetic airborne survey was conducted in 1957–9 for the Geological Survey and the United Kingdom Atomic Energy Authority.

Isotopic determinations of the ages of rocks of the Lizard were made by Miller and Green (1961) and Dodson (1961), and of granites and some other igneous rocks of south-west England by Miller and Mohr (1964); abstracts of these and other recent determinations are given by Sabine and Watson (1965; 1967; 1968). Green (1964a) considered the petrogenesis of the peridotite of the Lizard.

The geology of the peninsula has become an increasingly popular study of late and all these recent works furnish good bibliographies. The Ussher Society was founded in 1961, since when its proceedings have contained a number of useful papers, both specialist and general. In 1966 the Royal Geological Society of Cornwall published a commemorative volume on south-west England. The William Pengelly Cave Studies Association is furthering research into cave deposits.

The series of Geological Survey memoirs on Mineral Resources, in so far as they relate to Cornwall and Devon, is listed on p. 116, and information on metalliferous mining in south-west England has been collated by H. G. Dines (1956) and published in an exhaustive memoir.

In 1962 the newly formed South-West England field unit of the Geological Survey commenced six-inch mapping in the region; six one-inch sheets, covering most of the Carboniferous rocks, have now been completed and four descriptive memoirs written; this work is continuing.

2. Pre-Devonian Rocks

The open sea of the Armorican geosyncline of western Europe extended westwards from Poland across Germany, Belgium, northern France and southern England to south-west Ireland. Deposition of sediment on its bed continued from pre-Devonian times to the end of the Carboniferous Period. Probably this basin was an east–west branch of a major geosyncline which stretched from northern Norway to the southern United States (Fig. 4), Europe and North America having then not begun to drift apart, and ranged in time from late Pre-Cambrian to Carboniferous.

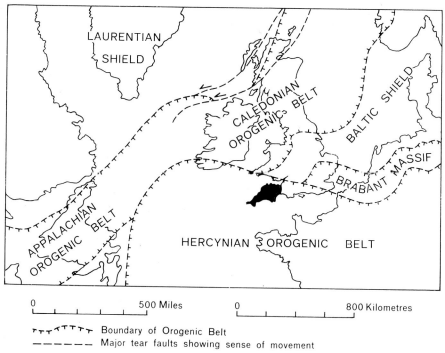

FIG. 4. *Geotectonic map of the north Atlantic region with Greenland and north-east Canada restored to their possible positions prior to continental drift. South-west England is shown in black*

The oldest rocks in and around south-west England are those of Brioverian (late Pre-Cambrian) age in Brittany and Normandy, the late Pre-Cambrian rocks of the St. David's area of South Wales and the probably Pre-Cambrian rocks of the Lizard and Eddystone Reef. Great orogenic movements, which may have ranged in time up to late Cambrian, imposed an east-north-east–west-south-west structural trend in northern France and South Wales and isotopic age determinations suggest that earth movements of similar age affected sediments in the Lizard area. At the end of Cambrian times a land mass of deformed rocks may have extended across the present south-west England (Fig. 5), forming the eroded metamorphic basement, now seen only at the Lizard and Eddystone, upon which rocks of later ages were deposited. The only known Ordovician and Silurian sediments of south-west England are exotic blocks of quartzite and limestone enclosed within Devonian strata, but sedimentation continued throughout these periods in both Wales and Brittany and similar conditions may have extended across the intervening region. A further major orogeny occurred at the end of Silurian times; it raised land to the north, and sediment-source studies in south Wales suggest that it may also have produced land areas in or a little to the south of the present south-west England.

Pre-Cambrian? Rocks of Eddystone Reef and the Lizard

Eddystone Reef is an isolated pinnacle of garnetiferous gneiss in the English Channel about 14 miles south of Plymouth; it is part of a major underwater outcrop of mica-schists and granitoid gneisses such as have not been found elsewhere in south-west England. Isotopic age determinations suggest they were last deformed towards the end of the Devonian period, but their highly metamorphosed state may indicate the effects of earlier orogenic movements, probably of Pre-Cambrian age.

The Lizard complex is faulted on its north side against Devonian sediments and forms one of the most interesting areas of south-west England (Fig. 7). Geophysical evidence suggests that these rocks do not extend laterally far beyond their known outcrop.

The geological history of the Lizard may have been as follows:

1. Deposition of muds and sandstones, volcanic tuffs and basaltic lavas, now present as the Old Lizard Head Series and Landewednack and Traboe Hornblende-Schists.
2. Local intrusion of acid sills, now the Man of War Gneiss.
3. Orogenic movements of Pre-Cambrian to Cambrian age, regional metamorphism and intrusion of peridotite, now altered to serpentine.
4. Intrusion of gabbro.
5. Orogenic movements, probably of Middle Devonian age, during which intrusion of basalt and dolerite dykes was closely followed by the intrusion of the microgranite of, for example, the Kennack Gneiss.
6. Orogenic movements of Permo-Carboniferous age which produced folding and faulting, including the possible thrusting or reverse faulting of the Lizard metamorphic complex against Devonian sediments.
7. Post-orogenic normal faulting.

All the older rocks of the Lizard have undergone dynamic, thermal or regional metamorphism which has produced a variety of metamorphic mineral assemblages. The Old Lizard Head Series, which is exposed on the west side of Lizard Point and north of St. Keverne, consists mainly of well-foliated mica-schists, quartz-granulites and hornblende-schists respectively formed from slates, sandstones and volcanic rocks. Andalusite, cordierite and anthophyllite-bearing rocks of the hornblende-hornfels thermal metamorphic facies are seen at Trelease Mill and Polkernogo, near St. Keverne, and staurolite, kyanite and sillimanite, almandine and epidote-bearing rocks of the almandine-amphibolite regional metamorphic facies at Polkernogo and at Pistil Ogo at Lizard Point.

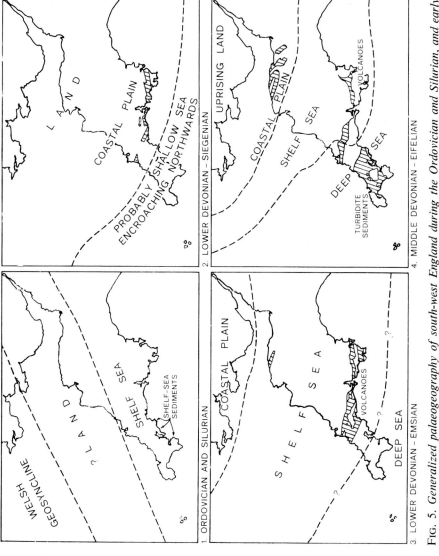

Fig. 5. *Generalized palaeogeography of south-west England during the Ordovician and Silurian, and early Devonian. Crosshatching indicates present-day outcrops*

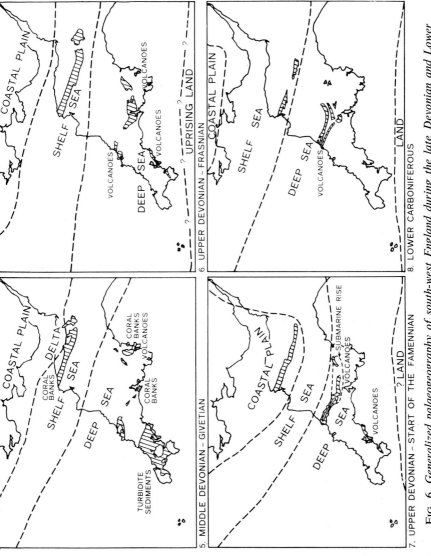

FIG. 6. *Generalized palaeogeography of south-west England during the late Devonian and Lower Carboniferous. Crosshatching indicates present-day outcrops*

In contrast, the Landewednack Hornblende-Schists form a monotonous and well-foliated sequence of hornblende and plagioclase-amphibolites with lenses and layers of epidote and grossular garnet, all of which appear to have been derived from a series of basaltic volcanic rocks and calcareous sediments. They were apparently interbedded with the Old Lizard Head Series and have undergone similar metamorphism. The Traboe Hornblende-Schists are Landewednack Hornblende-Schists which have undergone dynamothermal metamorphism during the intrusion of the peridotite, since wherever the intrusion of the peridotite into the Landewednack Hornblende-Schists can be identified a layer of Traboe Hornblende-Schists occurs at the contact, as between Carn Barrow and Kildown Point.

The Man of War Gneiss occurs at the southern end of Lizard Point, where it is exposed mainly in off-shore reefs and appears to have intruded the Old Lizard Head Series by lit-par-lit injection of quartz and feldspar. The multiple folding of the gneisses, similar to that of the enclosing rocks, suggests that they are an early, and probably pre-peridotite, intrusion.

The peridotite, now mainly altered to serpentine, occupies approximately 30 square miles of the Lizard peninsula and appears as a circular vertical plug-like mass with several small apophyses. Most of its boundaries are faulted, possibly as a result of contraction while cooling. Previously described as three ultrabasic intrusions, bastite-serpentine, tremolite-serpentine and dunite-serpentine, the peridotite is now (Green 1964) thought to comprise a single intrusion showing the effects of alteration. The primary peridotite (bastite-serpentine), a coarse-grained olivine-pyroxene-spinel assemblage, occurs mainly at the centre of the intrusive plug. Alteration has produced two mineral assemblages which together represent the tremolite-serpentine, the recrystallized anhydrous assemblage and the recrystallized hydrous assemblage. The former consists of fine-grained, banded olivine-pyroxene-plagioclase-spinel enclosing augen of primary pyroxene and porphyroblasts of spinel; it was probably formed during high-temperature dynamothermal alteration such as might occur in mylonitic zones of crushing. The second assemblage, fine-grained olivine-pargasite-pyroxene-spinel, appears to show further alteration and normally occurs at the boundary of the peridotite. The degree of serpentinization is variable; very finely recrystallized variants of both assemblages occur, usually completely serpentinized (dunite-serpentine).

The arcuate mass of gabbro around St. Keverne, with its abundant associated gabbro dykes, may form part of a ring complex intruded into the peridotite. Many unusual varieties of gabbro occur, especially in the dykes which have been metamorphosed by movements, probably of the dyke walls, at high temperature. This has produced augen, flaser and schlieren gabbros in which granulation of the original mineral fabric has taken place. Troctolite, a variety of gabbro consisting of olivine and plagioclase feldspar, occurs on the shore at and south of Coverack; it is intrusive into the serpentine and is itself cut by veins of gabbro.

The final important igneous event in the Lizard was the intrusion of basalt dykes and microgranite. These may have been associated with the general mid-Devonian orogenic movements which affected Wales and Brittany, but they are also possibly related to the abundant basic intrusions into the Devonian rocks of Devon and Cornwall. Basalt dykes, of feldspars with

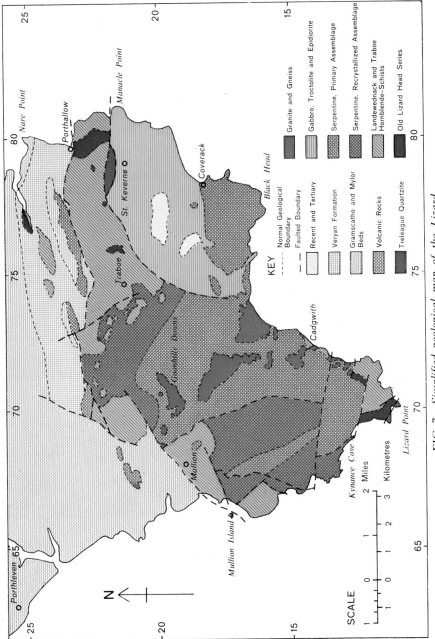

FIG. 7. *Simplified geological map of the Lizard*
After Geological Survey 1-Inch Sheet 359 and D. H. Green [1966]

KEY

Granite and Gneiss

Gabbro, Troctolite and Epidiorite

Serpentine, Primary Assemblage

Serpentine, Recrystallized Assemblage

Landewednack and Traboe
Hornblende–Schists

Old Lizard Head Series

Treleague Quartzite

Volcanic Rocks

Gramscatho and Mylor
Beds

Veryan Formation

Recent and Tertiary

Faulted Boundary

Normal Geological
Boundary

SCALE

Miles

Kilometres

Nare Point

Porthallow

Manacle Point

St. Keverne

Coverack

Black Head

Traboe

Cadgwith

Crousa Down

Lizard Point

Mullion

Kynance Cove

Mullion Island

Porthleven

N

olivine and pyroxene pseudomorphs in a fine-grained basalt matrix, and flat-lying basaltic sheets, now altered to actinolite-schists or amphibolites, occur as a north-north-west-trending swarm cutting the Lizard gabbro at Manacle Point and as a more irregular swarm cutting the serpentine and older rocks between Kennack Sands and Lizard Point. The younger acid intrusives crop out in the region of Kennack Sands and inland to the north-west. They comprise medium-grained non-sheared microgranite and contaminated acid injection-gneiss, only the former showing sharp intrusive boundaries. Field relationships show the acid gneisses to be the last important intrusions of the Lizard area.

During the Armorican orogeny the Lizard complex probably acted as a rigid metamorphic block subject only to large-scale faulting. Thrusting over the younger sediments to the north may have occurred, but the present northern edge of the complex is a zone of block faulting within which movements appear to have been mainly vertical. Similar faults are probably present at the northern boundary of the Eddystone Reef.

Ordovician and Silurian Rocks

Abundant exotic blocks of quartzite and limestone of Ordovician and Silurian age have been noted in the breccias and conglomerates of the Veryan Series (pp. 30–1). Their immense size, irregularity and prevalence at distinct horizons suggest they have not been transported far. Fossils obtained from them are mainly brachiopods, trilobites and corals. They include the following Middle Ordovician forms from quartzite near Perhaver: *Corineorthis decipiens*, '*Orthis*' *budleighensis*, *Orthambonites calligramma*, *Nesuretus sp.*, *Favosites sp.*, a Trinucleid, Cheirurid, Calymenid, Asaphid and a Phacopid; and the following probably Silurian fossils from limestone at Perhaver Beach and Catasuent Cove: *Phragmoceras sp.*, *Favosites fibrosus*, crinoid ossicles, *Scyphocrinites sp.* and Bryozoa.

The association of these blocks with coarse breccias and conglomerates, turbidite sandstones and spilitic volcanic rocks of Devonian age (pp. 26, 31) suggests unstable sedimentological conditions such as might mark a zone of submarine faulting at the junction of continent and geosyncline.

3. Devonian Rocks

Sedimentation in the sea of the Armorican geosyncline was greatest in Devonian times (Figs. 5 and 6). Environmental conditions varied and three main facies have been recognized in Europe. The Old Red Sandstone sediments are mainly fluviatile, lacustrine or deltaic. They comprise poorly sorted sandstones and conglomerates interbedded with marls and siltstones and are characterized by non-sequences, an ostracoderm-fish fauna and much debris of land plants. The near-shore marine facies consists of coarse-grained sandstones and siltstones in shales, with rare detrital limestones commonly composed of crinoid and thick-shelled brachiopod debris. Deep-water sediments comprise shales and limestones and contain corals, crinoids, cephalopods and eyed trilobites (Fig. 8).

Devonian? Rocks of Start Point and Dodman Point

The schists of Start Point are faulted against Devonian sediments to the north. They appear to form a westward-plunging anticlinorium between Salcombe and Bolt Tail and extend beneath the sea westwards to the West Rutts Reef and southwards for at least five miles to where they are overlain by New Red Sandstone rocks.

Two main rock types are distinguished: green (hornblende and chlorite) schists and mica-schists. In addition pink dolomitized limestone forms part of the East Rutts Reef. The green schists appear to be altered basic lavas or sills and comprise chlorite-epidote-albite-schists and hornblende-epidote-albite-schists according to the amount of dynamic metamorphism undergone; each type grades into the other.

The mica-schists are a uniform group composed of muscovite and quartz with albite and chlorite, while accessory minerals include tourmaline, epidote, rutile and ilmenite. They have formed from slate, siltstone and sandstone. The mica-schists and green schists are interbanded and evidently closely related in time the one to the other.

The schists are most likely altered Devonian rocks, but are possibly remnants of a Lower Palaeozoic or even pre-Palaeozoic land mass that has been brought into contact with Lower Devonian rocks by faulting. Recent structural studies suggest they have undergone only one major phase of deformation and their east–west-trending structures parallel those of the adjoining Devonian slates.

The rocks of Dodman Point may be faulted on the north, in this case against Devonian breccias and conglomerates. They closely resemble the Gramscatho Beds (pp. 26, 29) of the surrounding area in being composed of slightly metamorphosed subgreywackes and slates, and showing similar structural deformation.

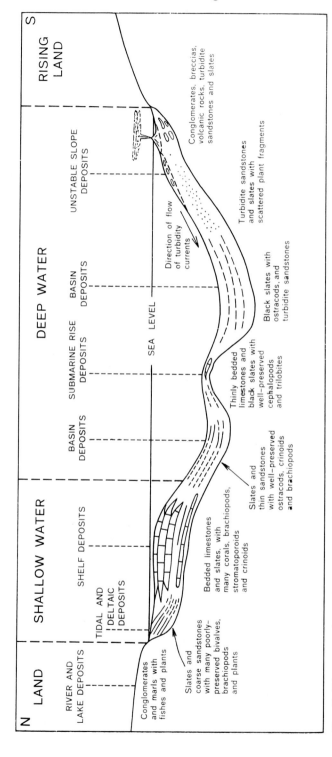

Fig. 8. *Diagrammatic section across south-west England during the Devonian Period showing the relations of sediments and fossils to their depths of deposition*

Lower Devonian

Lower Devonian times (Fig. 9) were marked by a passage from continental sedimentation to deposition in a shallow sea, which probably spread from the south-west, and by sporadic outbursts of vulcanicity. At the base of the succession in the south are the Dartmouth Slates (Table 1), of lower

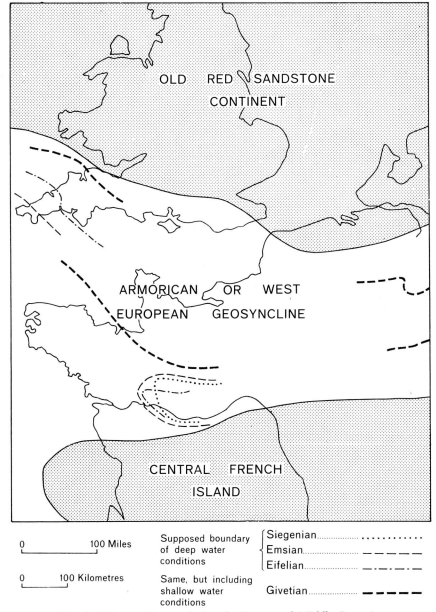

FIG. 9. *Western Europe during the Lower and Middle Devonian*
Based on H. K. Erben 1964

STAGES		1 NORTH DEVON	2 NORTH CORNWALL (TINTAGEL-NEWQUAY)	3 CENTRAL CORNWALL (LAUNCESTON-PENTEWAN)	4 CENTRAL DEVON (TAVISTOCK-BIGBURY BAY)	5 EAST DEVON (TORQUAY)	5 EAST DEVON (NEWTON ABBOT)	6 SOUTH CORNWALL (CHUDLEIGH)	6 SOUTH CORNWALL (PENTEWAN-MENEAGE)
UPPER	FAMENNIAN	Lower Pilton Beds / Baggy Beds / Upcott Beds / Pickwell Down Beds	Tredorn Slates / Woolgarden Slates / Delabole Slates	Stourscombe Beds	North Brenter Beds	Ostracod Slate of Anstey's Cove	Red Slaty Limestone	Mount-Pleasant Series	Veryan Series or Gidley Well Beds
	FRASNIAN	Morte Slates	Purple and Green Slates	Petherwin Beds	South Brenton Beds / Martavy Beds / Manor Hotel Beds / Whitchurch Green Slates	Slate with Volcanic Levels / Ostracod Slate of Saltern Cove		Kiln Wood Shales / Lower Dunscombe Goniatite Beds	Mylor and Gramscatho Beds
MIDDLE	GIVETIAN	Ilfracombe	Merope Island Beds	Slates with Volcanic Rocks and Calc-flintas	Ostracod Slate / Wearde Grit / Ostracod Slate	Saltern Cove Beds / Babbacombe Slates	Tuffs	Chudleigh Limestone	
	EIFELIAN	Jenny Start or Roadwater Limestone Beds	Longcarrow Cove Beds / Marble Cliff Beds		Warren Point Beds	Torquay Limestone	Limestone (including the Ashburton Limestone)		
	COUVINIAN		Slates of Trevone and Mother Ivey's Bay		Plymouth Limestone	Hope's Nose Beds / Shales with *Calceola*	Slate with fossils / Lava / Limestone		
LOWER	EMSIAN	Hangman Grits	Slates of Booby's Bay / Slates of Porthcothan / Slates of Bedruthan Steps	Staddon Grit		Meadfoot and Staddon Beds			
	SIEGENIAN	Lynton Beds		Meadfoot Beds					
	GEDINNIAN			Dartmouth Slates — Base not known					

TABLE 1. *Classification and stratigraphy of the Devonian rocks of south-west England*
Based mainly on M. R. House and E. B. Selwood [1966], with additions from
E. M. Lind Hendriks 1937, G. V. Middleton 1960 and B. D. Webby 1965

Map legend: Triassic and Permian / Carboniferous / Devonian / Granite. Scale: 0 — 50 Miles

Siegenian age, which were deposited in rivers, lakes and deltas. Their outcrop trends nearly due west from Dartmouth in Start Bay to Watergate Bay, north of Newquay, in the form of a major anticlinorium. The rocks are dominantly argillaceous and vary in colour from purple to green and grey. Some inter- bedded sandstones, conglomerates and pyroclastic rocks occur, especially in the eastern part of the outcrop. The following stratigraphical divisions and thicknesses of these beds in south Devon have been proposed (Dineley 1966):

Wembury Siltstones about 1000 ft	Shaly slates, siltstones, conglomerates and silty sandstones; red and green coloured; fossils numerous but fragmentary.
Scobbiscombe Sandstones about 400 ft	Sandstones and grits, massive and cross-bedded; pale grey; fossils at a few levels.
Yealm Formation (thickness unknown)	Slates, siltstones and sandstones with many pyro- clastic beds and conglomerates with volcanic debris; wide range of colours; fossils locally common.
Warren Sandstones about 300 ft	Sandstones and clay-slates finely interbedded; a few conglomerates; red and green coloured; fossils widespread and locally abundant. [Base not seen]

Similar lithologies can be traced westwards into the area of Polperro and Lantivet Bay in east Cornwall, though generally the Cornish succession is of finer grained sediments consisting of variegated purple and green slates interbedded with grey sandstones and with fish bone beds up to 3 ft thick. Interbedded green and purple slates are well exposed on the north Cornish coast at Watergate Bay. Deformed fossil fish remains abound in the bone beds and occur as isolated fragments in sandstones. Among the species recorded from south Devon are *Althaspis leachi, Rhinopteraspis cornubica, Europrotaspis sp., Drepanaspis carteri* and *D. edwardsii*, while east and north Cornwall have yielded many fragments of *Rhinopteraspis cornubica, Climatius sp.* and *Parexus sp.*

Around Stoke Climsland and the River Yealm in south Devon tuffs mark the beginning of vulcanicity which continued throughout Devonian and into Carboniferous times.

In south Devon and Cornwall the Dartmouth Slates are overlain con- formably by the Meadfoot Beds, a succession of slates, siltstones and sand- stones, with rare but persistent limestones, of marine origin. These sediments have been spasmodically subjected to vigorous current action, probably in shallow water. They show ripple-drift lamination, cross-lamination, a variety of erosional features such as scour and fill flute casts, intraformational conglomerates and erosion channels, and load casts, flame structures and ball and pillow structures (Plate 1A).

Rhythmic sedimentation is suggested by the finely alternating slates, silt- stones and sandstones, and many of the sandstones show graded bedding. The limestones are detrital, comprising fossil debris which originally accumu- lated as, for example, shell banks. In contrast, the calcareous sandstones and shales adjoining the limestones commonly contain well-preserved delicate fossil corals and spiriferid brachiopods. The Looe Grits, which contain the richest fossiliferous horizons of the Meadfoot Beds, have been traced both west and east from their main outcrop between Looe and Fowey, and similar lithologies can be seen at Newquay and Tor Bay.

Fossils from the Meadfoot Beds include '*Rhynchonella*' *pengelliana, Spirifer paradoxus, Stropheodonta gigas, Chonetes sarcinulatus, Acrospirifer pellico, Pleurodictyum problematicum, Thamnopora cervicornis, Syringaxon sp., Burmeisterella elongata* and *Digonus goniopygaeus.*

Volcanic activity during deposition of the Meadfoot Beds resulted in the formation of coarse agglomerates and tuffs; these are well exposed in the coast south of Brixham and in St. Austell Bay.

The Staddon Grits appear to follow in normal succession above the Meadfoot Beds but are probably in part of equivalent age. They consist of sandstones, intraformational conglomerates and thin limestones of Emsian age.

In north Devon and Somerset (Fig. 10) the oldest Devonian rocks are the Lynton Beds, of late Emsian–Eifelian age, which are bioturbated, fine-grained, laminated sandstones and mudstones with thin shell beds. These sediments are about 1300 ft thick; they accumulated rapidly in shallow water and were occasionally subjected to penecontemporaneous erosion. Thus their depositional environment resembled that of the Meadfoot Beds to the south.

Bivalves predominate in the rather sparse fossil fauna of the Lynton Beds— *Modiomorpha sp., Pterinea sp., Limoptera sp.* and *Actinodesma sp.* Other fossils include *Platyorthis circularis, Spirifer subcuspidatus,* fenestellid bryozoans and possible Pteraspids, and the trace-fossil *Chondrites* is ubiquitous.

Middle Devonian

Sea covered Devon and Cornwall during Middle Devonian times but the presence of land to north and south resulted in variations in facies from deep water to fluviatile which continued throughout the Upper Devonian (Fig. 11). Continuous sedimentation was maintained in central and north Devon and probably also in south-west Cornwall, where the Gramscatho Beds were being deposited.

The almost unfossiliferous sediments north of the Lizard in south-west Cornwall are now generally agreed to be probably of Middle or Upper Devonian age. The following rock groups and probable ages are recognized:

Veryan Series or Gidley Well Beds   Upper Devonian?, pp. 30–1
Mylor Beds   ⎫
Gramscatho Beds   ⎬   Middle Devonian?
                            ⎭

The Gramscatho Beds are formed of interbedded greywackes and slates with sporadic limestones, conglomerates, cherts and spilitic lavas, and the Mylor Beds comprise slates and siltstones with rare sandstones. All these rocks have undergone repeated folding and faulting and the formation boundaries, especially that between the Gramscatho Beds and Mylor Beds, are ill defined. Structural studies in the Porthleven area indicate that the Mylor Beds are younger than the Gramscatho Beds. Fossils from the Gramscatho Beds, and in particular *Dadoxylon hendriksi*, which is related to

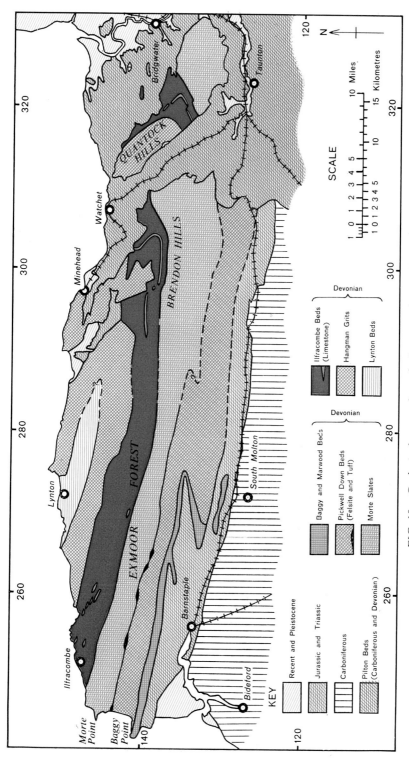

FIG. 10. *Geological map of north Devon and west Somerset*
After W. A. E. Ussher 1907, J. G. Hamling 1910 and B. D. Webby 1965

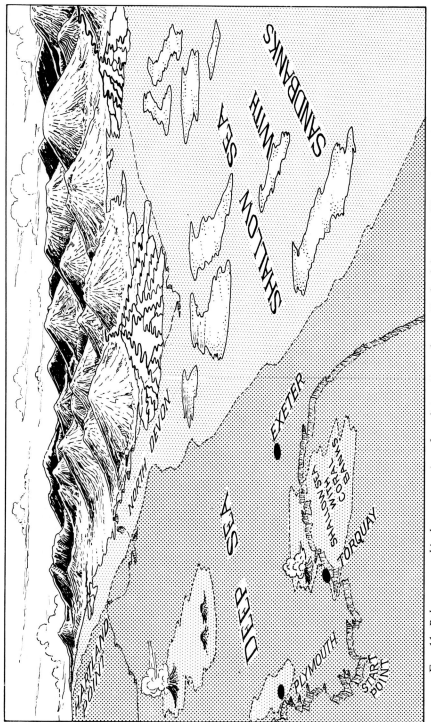

FIG. 11. *Palaeogeographical reconstruction of conditions in south-west England during Middle and Upper Devonian times*

a variety of wood known from the Lower and Middle Devonian rocks of central Europe, indicate a probably Middle Devonian age. At Pentewan the Gramscatho Beds appear to lie with slight unconformity on the Lower Devonian Meadfoot Beds.

The accumulation of the thick geosynclinal flysch-type sediments of south-west Cornwall, now the greywackes and slates of the Gramscatho Beds and Mylor Beds, took place in a rapidly subsiding east–west trough. At the end of Middle Devonian times the area was apparently affected by earth movements occurring to the south, and considerable outpouring of spilitic lava took place possibly combined with low-grade regional metamorphism.

The Middle Devonian sediments between Padstow and Torquay indicate that uniform deeper-water conditions had become established in this area by late Middle Devonian (Givetian) times. The arenaceous Staddon Grits are succeeded by lower Eifelian black slates, with some volcanic lavas and tuffs, which have yielded *Nereitopsis mira, Pleurodictyum* and solitary corals such as Zaphrentids and *Calceola sandalina*, and brachiopods, trilobites and the characteristic goniatite *Anarcestes lateseptatus*. Many of these delicate fossils are well preserved, even though distorted by subsequent orogenic movements, and must have lived and died in quiet water.

In north Cornwall this sequence of slates with volcanic rocks contains only scattered thin limestones but around Plymouth, Torquay and Chudleigh slates commonly give way to thick limestones. These major limestones are broadly grouped as massive or thinly bedded. The former are composed either of abundant massive stromatoporoid colonies with which grew a few corals, as in the Walls Hill limestone, Torquay, or mainly of rugose and tabulate coral colonies with some stromatoporoids, as at the Richmond Walk Quarry, Plymouth. Wolborough Quarry and Barton Quarry, and in particular the exposure of the Lummaton Shell Bed at Lummaton Quarry, have yielded a rich and varied fauna including the Middle Devonian brachiopods *Stringocephalus burtini, Hypothyridina cuboides* and *Mimatrypa desquamata*. The limestones are generally pale grey and very pure; they have been extensively quarried for decorative purposes and may be seen in the entrance hall of the G.P.O. Tower, London. The thinly bedded limestones are usually dark grey and vary from calcareous muds containing unbroken fossils to completely fragmental limestones. Some may be of chemical origin. They were probably formed in shallow water receiving only a little non-carbonate sediment, rather like the present-day tidal zone around the flat banks of the Bahamas. Dyer's Quarry, Torquay, at the western end of the Daddy Hole limestone mass, has a rich coral fauna preserved in thinly bedded, dark grey limestones exposed at the base of the quarry. *Thamnophyllum germanicum schouppei* and *Mesophyllum maximum* are abundant and appear to be preserved in their positions of growth.

Typical Middle Devonian fossils found in both limestone groups include: *Acanthophyllum spp.*, including *A. heterophyllum, Grypophyllum tenue, Mesophyllum spp., Phillipsastrea hennahii hennahii, Thamnopora sp., Alveolites sp., Coenites sp., Stromatopora concentrica, Actinostroma clathratum* and *Amphipora ramosa*.

The volcanoes of Middle Devonian times in south-east Devon probably assisted the formation of these limestones by building shallow sea floors on which the colonies grew and then subsequently emitted poisonous gases which killed the polyps.

In north Devon and Somerset the Middle Devonian strata originated in shallow water near the shore of the still-emerging Old Red Sandstone continent to the north. The Lynton Beds are succeeded by the fluviatile and deltaic Hangman Grits, which are about 5000 ft thick and have been divided into:

'*Stringocephalus*' Beds ⎫
Sherrycombe Beds   ⎬ Upper Hangman Grits
Rawn's Beds   ⎭
Trentishoe Grits

These sediments vary from clay-slate to conglomerate and show ripple-marks, convolute bedding and channelling. Fossils from the Sherrycombe Beds and '*Stringocephalus*' Beds are mainly bivalves such as *Myalina*, with a few corals, brachiopods and gastropods. These last suggest the beginning of a renewed north-eastward marine transgression across north Devon and Somerset, which in the Quantock Hills area continued into early Ilfracombe Beds times (Fig. 6).

The Ilfracombe Beds, whose deposition extended from upper Middle Devonian into Upper Devonian times, comprise marine slates with thin fossiliferous limestones, and also sandstones and slates probably of shallow water or deltaic origin. Near Ilfracombe the middle Ilfracombe Beds contain four limestones, in ascending order the Rillage Limestone ($\frac{1}{2}$ to 2 ft thick), the Jenny Start (or Roadwater) Limestone (about 30 ft), the Combe Martin Beach Limestone (3 to 5 ft), and the David's Stone Limestone (about 30 ft). Each contains a distinct coral fauna and ages range from Givetian to Frasnian. The corals include *Disphyllum, Thamnophyllum, Endophyllum, Acanthophyllum, Mesophyllum, Phillipsastrea, Metriophyllum, Syringaxon, Thamnopora, Favosites* and *Alveolites*. Some are in their positions of growth, enclosed in stromatoporoids, as in the David's Stone and Jenny Start limestones, but some limestones are entirely detrital and packed with abraded fossil fragments, largely of bryozoa, brachiopods, crinoids, gastropods and algae, as for example the Rillage and Combe Martin Beach limestones. The Roadwater Limestone forms a useful marker horizon that has been traced from Ilfracombe to the Quantock Hills.

Upper Devonian

Upper Frasnian times saw the start of the Upper Devonian–Lower Carboniferous marine transgression northwards into Wales and central England. The uprising Bretonic land areas to the south, possibly the southern boundary of the geosyncline, provided coarse debris for conglomeratic sedimentation in south Cornwall. In south Devon the massive mainly Middle Devonian limestones were succeeded by deeper water cephalopod limestones and ostracod shales, while in north Devon and Somerset the fluctuating northern boundary of the geosyncline resulted in deposition of sediments ranging from shallow marine to deltaic and fluviatile.

In south-west Cornwall the Veryan Series crops out from Gorran Haven to Mullion and is faulted against or unconformable on the Lizard rocks and the

(MN 23441)

A. Meadfoot Beds, Lantic Bay, Fowey

Plate 1

(*For full explanation see p. ix*)

B. Veryan Series unconformable on Middle Devonian? slates,
Greeb Point, St. Gorran, Cornwall

(MN 23442)

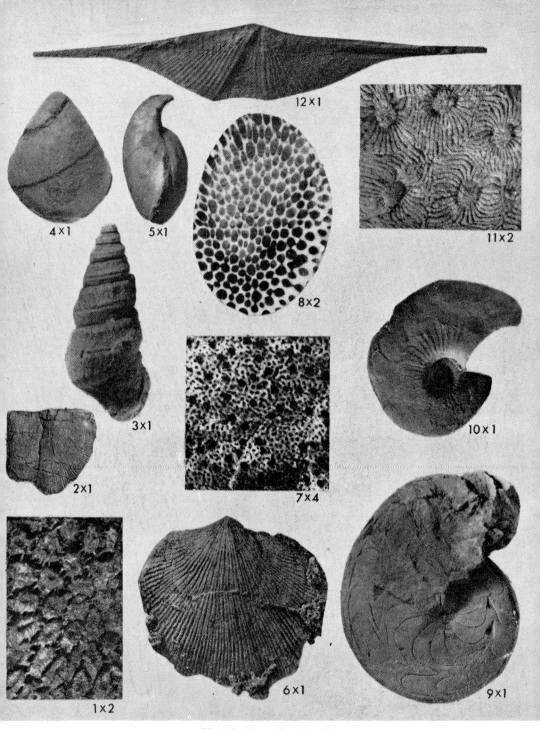

Plate 2: Devonian Fossils

1. *Pleurodictyum sp.*—Lower Devonian. **2.** *Douvillina elegans* (Drevermann)—Lower Devonian.
3. '*Murchisonia*' *bilineata* (Dechen)—Middle Devonian. **4, 5.** *Stringocephalus burtini* (Defrance)—
Middle Devonian. **6.** *Mimatrypa desquamata* (J. de C. Sowerby)—Middle Devonian. **7.** *Stromatopora sp.*
with *Syringopora sp.* (formerly '*Caunopora placenta*')—Middle Devonian. **8.** *Thamnopora cervicornis*
(Blainville)—Middle Devonian. **9.** *Sporadoceras orbiculare* (Münster) var.—Upper Devonian. **10.** *Cymaclymenia cordata* Wedekind—Upper Devonian. **11.** *Phillipsastrea hennahii* (Lonsdale)—Upper Devonian.
12. *Cyrtospirifer verneuili* (Murchison)—Upper Devonian.

Middle Devonian Gramscatho Beds. It comprises breccias and conglomerates interbedded with greywackes, slates and limestones (Plate 1B), probably derived from the land areas to the south. Ordovician quartzite and Silurian, Siegenian, Emsian and Middle Devonian limestone fragments have been identified (p. 20). Limestones occurring with the pillow lavas of Mullion Island have yielded Frasnian conodonts.

In the Port Isaac–Tor Bay region sedimentation continued without change from the Givetian into the Frasnian but then soon gave way to deposition in fairly well-defined areas marked by three main lithologies: ostracod slates with abundant volcanic rocks; cephalopod limestones and slates; brachiopod-crinoidal slates.

The ostracod slates crop out mainly between Padstow and Torquay and represent an eastward extension of the deep-water sedimentation already established in the Padstow area in Middle Devonian times. The cephalopod limestones and slates occur mainly around Tavistock and Chudleigh, and were probably formed on submarine ridges. Fossils of the brachiopod-crinoidal slates, which occur between Launceston and Tintagel, are generally very well preserved, indicating a still-water environment—probably a submarine shelf with little current action.

In the Padstow area Upper Devonian times began with large-scale vulcanicity during which the 200-ft-thick pillow lavas of Pentire Point were formed. The possibly equivalent rocks south of Padstow, the Longcarrow Cove Tuff Beds, consist of grey slates with interbedded tuffs and agglomerates. Above lie the Lower Merope Island Beds, banded dark grey slates at least 150 ft thick, which contain the small pyritized goniatites *Manticoceras*, *Aulatornoceras*, *Ponticeras*, *Archoceras* and *Tornoceras* of middle Frasnian age, and the trilobite *Ductina ductifrons*. The overlying banded slates, the Purple and Green Slates, are at least 400 ft thick; they are the highest horizon seen here and occupy the core of the St. Minver synclinorium. These slates are rich in ostracods, including *Richterina* and *Entomis serratostriata*; other fossils include *Posidonia venusta*, and *Ductina ductifrons*, while the middle Frasnian goniatites *Subtornoceras*, *Tornoceras* and *Manticoceras* occur near the base in the Gravel Caverns Conglomerate.

In south Devon the massive limestones around Plymouth and Torquay are overlain by slates with some volcanic rocks and thin, usually lenticular, limestones. Near Torquay the massive limestones contain lower Frasnian fossils including the corals *Frechastraea spp.* and *Marisastrum marmini*. At Plymouth purple and green banded slates have yielded *Manticoceras* and ostracods. Ostracod slates between Plymouth and Dartington contain turbidite sandstones such as the Wearde Grit. At Dartington and around Torquay pillow lavas and tuffs alternate with ostracod slates and, south of Newton Abbot, stromatoporoidal limestones. These limestones probably formed on local shallower sea floors built up by the volcanic activity. The Dartington volcanics, mainly keratophyric tuffs, are interbedded with slates that have yielded the goniatites *Tornoceras simplex*, *T. undulatum* and *Gephyroceras sp.* Slates on the promontory in Saltern Cove near Torquay have yielded upper Frasnian goniatites including *Archoceras angulatum*, *A. varicosum*, *A. ussheri*, *Manticoceras cordatum*, *Crickites holzapfeli*, *Tornoceras* and *Aulatornoceras*. At Anstey's Cove *Richterina* (*Maternella*) *hemispherica, R. (M.)* cf. *dichotoma* and *R. (R.) striatula* occur in Famennian slates with volcanic beds, and fossils from Knowles Hill, north of Newton Abbot, include abundant remains of the trilobites *Trimerocephalus mastophthalmus* and *T. trinucleus.*

The best development of the cephalopod limestones and slates is in the Chudleigh district, where Frasnian fossils, including *Manticoceras, Beloceras* and *Tornoceras,* have been obtained from the Lower Dunscombe Goniatite Bed immediately above the Chudleigh limestone. At Mount Pleasant the whole of the Famennian sequence is condensed into 170 ft of dark grey slates with calcareous nodules; the fauna includes *Parawocklumeria, Kalloclymenia, Cyrtoclymenia, Cymaclymenia, Richterina (Richterina)* and *R. (Maternella).* Cephalopod-rich limestones and slates also occur between Tavistock and Okehampton and in the Launceston district, but in both areas they contain interbedded brachiopod-crinoid slates and some ostracod slates. Thus the Upper Devonian slates of west Dartmoor contain the brachiopod *Ambocoelia urei,* Spiriferids and the bivalves *Myalina, Paracyclus, Posidonia* and *Sanguinolaria;* the goniatites *Imitoceras* cf. *sulcatum, Platyclymenia (Pleuroclymenia?) sp., Cymaclymenia* and *Sporadoceras* and the trilobite *Phacops (Cryphops?) wocklumeriae.* Around Launceston a sporadically fossiliferous and partially metamorphosed succession of slates and calcflintas, probably of Upper Devonian age, is succeeded by the Petherwin Beds and the Stourscombe Beds. Both these formations are of upper Famennian age and contain lenticular limestones which are locally packed with goniatites, the Stourscombe Beds also occurring as black slates containing a well-preserved fauna of trilobites, brachiopods, ostracods and goniatites. The Lower Petherwin Beds have yielded *Gonioclymenia, Clymenia, Kosmoclymenia, Cymaclymenia, Costaclymenia, Platyclymenia, Imitoceras, Sporadoceras, Phacops (Phacops) granulatus, P. (P.) accipitrinus accipitrinus* and *Cyrtosymbole (Waribole?) dunhevedensis,* and the Stourscombe Beds have in addition yielded *Kalloclymenia, Wocklumeria, Epiwocklumeria, Postglatziella, Parawocklumeria, Kenseyoceras, Discoclymenia, Phacops (P.), Phacops (Cryphops?), P. (Dianops)* and *Chaunoproetus.*

The Upper Devonian rocks between Launceston and Tintagel comprise fine-grained, homogeneous smooth and well-cleaved grey slates which have been extensively quarried between Delabole and Tintagel. Few limestones and sandstones occur, but thin limestones rich in brachiopods, particularly *Cyrtospirifer verneuili,* bryozoa, bivalves and crinoids have been noted at Trebarwith Strand and Tregardock. Slates in the Old Delabole Quarry and elsewhere have yielded flattened fossils, especially the 'Delabole butterfly', *C. verneuili,* and complete crinoids. The good state of fossil preservation and the almost complete absence of current structures, apart from those in sandstones near Trebarwith Strand, suggests that these slates were deposited in quiet water carrying very little sediment.

In north Devon and Somerset the upper Ilfracombe Beds are of Frasnian age and comprise a shallow-water succession of shales, siltstones and thin sandstones. Thin discontinuous limestones occur in the Brendon Hills and the Quantocks; they are formed largely of crinoidal debris, which suggests that the waters had become unsuitable for the growth of corals. The Morte Slates appear to lie conformably on the Ilfracombe Beds and yield a rather sparse fauna which includes poorly preserved *Cyrtospirifer verneuili* of Upper Devonian age. They comprise about 5000 ft of smooth and glossy grey or purple slates with scattered calcareous nodules and thin sandstones of shallow marine origin. In the Brendon Hills the middle of the succession shows cross-laminated sandstones and siltstones which appear to have been deposited at a delta front.

The base of the succeeding Pickwell Down Beds, which are of middle Famennian age, is marked by a tuff band, the Bittadon 'Felsite', which has yielded fragments of armoured fish characteristic of the Upper Devonian, for example *Holonema* cf. *ornatum, Bothriolepis, Holoptychius, Polyplocodus* and *Coccosteus*. Above the tuff lie nearly 4000 ft of red, purple, brown and green sandstones, with some bands of greyish blue shales, marking an interval of continental fluviatile sedimentation.

There followed a gradual subsidence in north Devon. The Upcott Beds comprise about 800 ft of yellow and green cleaved sandstones and slates which accumulated in marshy lagoons where fossils were not preserved. The succeeding Baggy Beds are well exposed at Baggy Point as massive cross-bedded and more thinly bedded sandstones and siltstones, intraformational conglomerates, slumped sediments and thin crinoidal and gastropodal lime-stones with *Lingula*. The surfaces of the sandstones commonly show evidence of contemporaneous wave action and are usually bioturbated with abundant trace fossils such as *Teichichnus* cf. *rectus, Monocraterion* cf. *tentaculatum, Arenicolites curvatus* and *Diplocraterion yoyo*. Inland the Baggy Beds at Marwood and at Sloley Quarry have yielded a good fossil flora, including *Sphenopteridium rigidum, Xenotheca devonica, Sphenopteris, Telangium, Knorria* and *Cordaites*. These sediments were deposited largely on a delta-front platform. They are about 1500 ft thick.

Deltaic conditions continued into early Pilton Beds times, as evidenced by the occurrence of *Nudirostra laticosta*, but gave way to a shallow-water, near-shore, marine environment by the start of the Carboniferous. The topmost Devonian strata of north Devon, the lower Pilton Beds, consist of fossiliferous thinly bedded bluish grey slates and limestones with current-bedded sandstones which become increasingly cherty upwards. Their fauna is equivalent to that of the Etroeungt Beds of the Ardennes and includes *Avonia bassa, Productella (Hamlingella) goergesi, P. (H.) piltonensis, P. (Whidbornella) caperata, P. (W.) pauli, P. (Steinhagella) steinhagei, Athyris concentrica, Cyrtospirifer verneuili guttata*, the trilobite *Phacops (Phacops) accipitrinus accipitrinus* and the conodonts *Gnathodus* and *Spathognathodus*.

The Stourscombe Beds, Yeolmbridge Beds (p. 35), Meldon Slate-with-lenticles Formation (p. 37) and Pilton Beds are best regarded as a single Transition Group spanning the Devonian–Carboniferous junction.

4. Carboniferous Rocks

The Carboniferous rocks of south-west England have long been known as the Culm Measures, possibly on account of the occurrence at a few localities, mainly in the Upper Carboniferous, of soft sooty coal which in Devon is known as 'culm'; however, the word culm may be derived from the Old English col (coal) or the Welsh cwlwm (knot), a reference to the contortions common in the beds. The Culm Measures occupy an area of more than 1200 square miles. Their outcrop in the north of the region extends eastwards from near Barnstaple to within about four miles of Wellington in Somerset, and in the south from Boscastle to the east side of Dartmoor (Fig. 1). The northern outcrop of the basal beds, although faulted in places and faulted out between South Molton and Brushford, is fairly continuous and there are no inliers of Devonian rocks. In the south, however, their outcrop is much broken by thrusts and faults, and there are inliers of Upper Devonian rocks, while outliers of Culm beds are found on the Devonian formation; a narrow outcrop of the Lower Carboniferous may be traced from the coast near Boscastle to skirt the northern edges of Bodmin Moor and Dartmoor.

The measures are divisible into three. The Lower Carboniferous forms only narrow belts on both northern and southern flanks of the great central area occupied by the Upper Carboniferous. It comprises dark grey shales with thin sandstones and local slates, lavas, tuffs, agglomerates, limestones and cherts. Although the limestones are lenticular, the Lower Carboniferous sediments in general are of fairly uniform lithology and probably accumulated slowly under quiescent conditions disturbed only by periodic vulcanicity; the radiolarian cherts may have formed in lagoons. The Crackington Formation, at the base of the Upper Carboniferous, consists principally of shales and thin turbidite sandstones. It is now recognized as including the Welcombe Formation, previously thought to be topmost Carboniferous. The overlying Bude Formation consists of massive sandstones with shales.

An original three-fold division of the Culm was proposed by Ussher in 1887 and modified by him in 1892 and 1900. It superseded the two divisions of Sedgwick and Murchison (1840), although this two-fold classification was adopted by Dearman in 1959. Table 2 compares Ussher's lithological groups with more recently defined ones and with the classification now adopted by the Geological Survey, whose palaeontological evidence suggests that the Lower Carboniferous and the Crackington Formation correspond approximately to the Dinantian (Tournaisian and Viséan) and Namurian series of the Carboniferous, and the Bude Formation to the lower part of the Westphalian Series (Edmonds 1974).

Near Chudleigh and Newton Abbot conglomeratic sandstones (Ussher's Ugbrooke Park Beds), possibly of Crackington Formation age, rest on Lower Carboniferous cherts. Several authors have regarded the junction as an unconformity. However, Butcher and House (1972) showed by trenching that on the Chudleigh escarpment *Posidonia* Beds of the Lower Carboniferous were overlain conformably by conglomeratic sandstones.

Oolitic and crinoidal limestones of a small Carboniferous inlier at Cannington Park, north-west of Bridgwater, have been assigned to the Upper *Caninia* Zone of the Carboniferous Limestone.

Overfolding is common in the Lower Carboniferous and Crackington Formation, the folds generally being overturned to the north on the northern side of the synclinorium and to the south on the southern side. Fold attitudes change eastwards in the southern crop of these beds and near-horizontal axial planes in the north Cornwall cliffs (Plate 4B) contrast with those dipping about 45° N.N.W. in the Okehampton area. Intrusion of granite has locally pushed the folds into a more upright position and in places, for example the country bordering the north-eastern edge of the Dartmoor Granite, extensive development of faults on the steeper bedding planes has cut out many normal fold limbs. Folding in the more competent Bude Formation, and in the Crackington (Welcombe) Formation near the centre of the synclinorium, is generally open, with overturned strata being mainly confined to more intensely disturbed areas adjoining faults.

Lower Carboniferous

Cornwall and South Devon

In Cornwall the Lower Carboniferous extends from the coast south of Boscastle to Launceston and thence forms an arc around the north of Dartmoor to Drewsteignton (Plate 4A). It flanks the eastern edge of the granite southwards from Christow to Cornwood. A second outcrop to the south of the first strikes generally parallel to the northern crop and extends eastwards to the granite at Lydford. Lower Carboniferous strata also occur between Tavistock and Marytavy. The outlier forming the hilly ground around St. Mellion and St. Dominick may comprise both Lower Carboniferous and Crackington Formation rocks.

The order of succession of the beds is commonly difficult to determine on account of overthrusts and inversions. However, Selwood (1960; 1961) defined the basal beds of the Carboniferous at Boscastle as a group of dark grey phyllites containing trilobites of the *Gattendorfia* Zone, and in the Launceston area as the Yeolmbridge Beds, greyish brown silty slates with a few lenticular limestones containing goniatites and trilobites of this zone and overlying the Stourscombe Beds which yield goniatites and trilobites of *Wocklumeria* (topmost Devonian) age. He recorded that at Gull Rock in the coastal area near Buckator Cliff thin limestones (Buckator limestones) within a group of slates yielded goniatites tentatively referred to the P_1 Stage. Two thin lenticular limestones of the Buckator Formation at this locality have yielded mid-Dinantian conodonts and a Tournaisian coral and brachiopod fauna has been found in the lower of these limestones. The basal Carboniferous black shales and fine sandstones are faulted against Upper Devonian slates on the coast south of Boscastle. The Barras Nose Formation (Plate 3A), a sequence of shales and siltstones with thin sandstones and with crinoidal limestones towards the top, is exposed at Willapark and at Tregardock south-west of Boscastle. It is overlain by tuffs, agglomerates and lavas of the Tintagel Volcanic Formation, exposed in the cliffs at Boscastle, and succeeding slates of the Trambley Cove Formation, the three divisions forming the

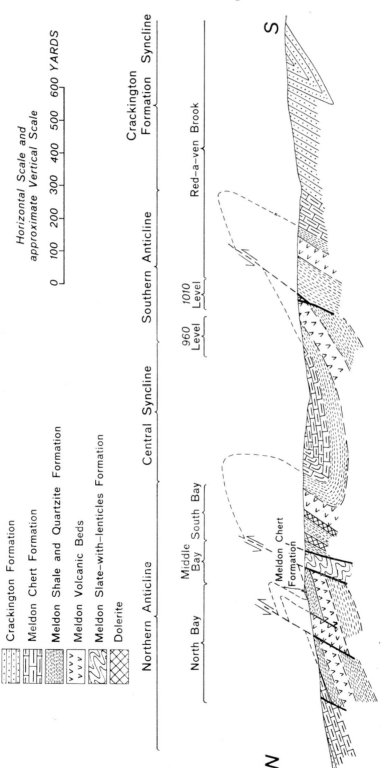

FIG. 12. *Section across the Meldon Quarry area*

From 'Geology of the country around Okehampton', *Mem. Geol. Surv.* 1968

Tintagel Group. Conodonts from the Barras Nose Formation show some overlap in age with those from the Buckator limestones but are also of mid-Dinantian age and both formations are younger than the Yeolmbridge Beds. Strong beds of Lower Carboniferous chert form the outstanding eminence of Fire Beacon Point and are probably younger than the Buckator and Barras Nose formations. Beds of chert stretch discontinuously across the country in a series of whale-back ridges, arranged *en échelon*, past Otterham, Tregeare, Launceston, Lifton, Bridestowe and Meldon to Sticklepath and Drewsteignton. They are probably the most easily mapped of the Lower Carboniferous formations and locally contain lenticular black limestones which have in the past been worked to their lateral limits and to a depth governed by the cost of raising the stone and dewatering the pit; such diggings are now commonly represented by deep pools concealed by wooded slopes, as at Sourton, Meldon and South Tawton. The limestones and associated shales commonly yield goniatites and *Posidonia* of Lower Carboniferous age, and in the Okehampton district the P_{1c}, P_{1d} and P_{2a} zones have been recognized. In the Launceston area Cannapark Quarry has yielded *Goniatites crenistria* indicating Zone P_{1a}. Shaly and lenticular limestones at Halton Quay, on the Tamar estuary, have yielded *Posidonia becheri* and foraminifera comparable with some found at Westleigh in north Devon.

Rocks of Lower Carboniferous age on the coast at Boscastle lie in several tectonic slices separated by thrusts or faults. The cliffs between Boscastle and Fire Beacon Point reveal many folds with nearly horizontal axial planes, the curved beds of sandstone being described locally as fossil trees. The contorted sandstones, slates, limestones and cherts show superimposed folding and are faulted against intensely contorted fossiliferous Crackington Formation sandstones and shales at Rusey. The beds cannot yet be definitely correlated with the sequence inland near Okehampton, where extensive quarrying of Carboniferous rocks within the Dartmoor Granite aureole at Meldon by British Rail has shown the Lower Carboniferous to be disposed in two anticlines overturned to the south-east (Fig. 12). The core of the northern anticline comprises tightly folded slaty and shaly hornfels with thin bands and streaks of silty quartzite. Dearman and Butcher (1959) regarded this Slate-with-lenticles Formation, mainly on structural evidence of two ages of folding, as spanning the junction between Upper Devonian and Lower Carboniferous, but the structure may be the result of crumpling in an anticlinal core. Dearman (1959) showed that the overlying Lower Carboniferous rocks comprised 85 ft of black chiastolite slate passing up into inter-bedded shale and quartzite 175 ft, overlying which were tuffs with inter-bedded shale and quartzite 85 to 200 ft, black shale with quartzite lenticles 70 ft, massive black limestone with subordinate chert, siliceous mudstone and shale 102 ft (Plate 3B), thinly bedded shale with thicker chert beds 120 ft, and thin black limestones with shale, siliceous mudstone and chert 18 ft. The volcanic rocks may equate with the Tintagel Volcanic Formation and the top 240 feet, of shale, chert and limestone, with the Fire Beacon Chert Formation. Elsewhere on the northern and western margins of Dartmoor the Lower Carboniferous rocks (Plate 4A) lie in a single overturned anticline. Between Meldon and Tavistock Dearman and Butcher found beds similar to the Slate-with-lenticles Formation overlying their North Brentor Beds (mainly slates) of possible *Wocklumeria* age; stratigraphically beneath lay the South Brentor Beds (slate, shale and calcareous siltstone) and Marytavy Beds (slates) of possible *Clymenia* age, and Manor Hotel Beds (mainly slates) of *Platyclymenia* age.

East of Dartmoor, around Chudleigh, siliceous shales with limestone nodules containing ammonoids of *Wocklumeria* age are overlain by siliceous shales containing conodonts correlated with the *Pericyclus* Zone at the top of the Tournaisian (House and Butcher 1962), and south of Ugbrooke Park shales with goniatites referable to the P_{2c} Zone at the top of the Viséan have been described by Butcher and Hodson (1960). Successions are summarized in Table 2. Good fossil localities include Waddon Barton and Hestow Farm, Ideford. The widely developed horizon of *Neoglyphioceras spirale* is present at Waddon Barton together with *Mesoglyphioceras [Goniatites]* aff. *granosum* and *Posidonia becheri* and this may indicate a low P_2 age.

Volcanic activity was widespread during Lower Carboniferous times and extensive sheets of spilitic lava, tuff and agglomerate occur in north Cornwall and south Devon. These lavas are commonly sheared but locally pillow forms have been preserved. The bold eminence of Brent Tor consists almost entirely of 200 ft of spilite, commonly brecciated but where locally better preserved, amygdaloidal and pumiceous. Manganese ores have been worked in the lava here and at Lewannick. Similar lava masses form much of the picturesque land near Launceston, and Sourton Tors are composed of massive tuff and agglomerate.

North Devon

Goldring (1955) confirmed that in north-west Devon the base of the Carboniferous occurs in the Pilton Beds, a sequence of shale and slate with thin bands of sandstone and limestone, within which beds containing *Phacops* and referred to the *Wocklumeria* Zone are overlain by beds containing trilobites of the *Gattendorfia* Zone; few goniatites have been found. Thus the base of these beds lies near the stratigraphical position suggested by Dearman and Butcher (1959) for the Slate-with-lenticles Formation in the south. The topmost Pilton Beds in north-west Devon are a series of unfossiliferous shales. In the north-east of the county the probably equivalent beds are hard black slates; no fossils have yet been found in them.

As in Cornwall and south Devon, a series of whale-back chert ridges extends eastwards from Codden Hill through Hangman's Hill and by Swimbridge towards South Molton. Again as in the south, black limestones associated with the cherts contain *Posidonia*. Goniatites of upper Viséan age occur in these rocks and associated shales near Fremington and at Venn and Swimbridge, near Barnstaple, and also at Bampton. A study of corals and brachiopods from Codden Hill led Vaughan (1904) to compare these beds with those of the *Zaphrentis* Zone of the Carboniferous Limestone near Bristol, and to suggest a break in sedimentation within the Lower Carboniferous. However, no further evidence has been discovered to support this view. Prentice (1960a) noted that the top of his Chert Beds in the Barnstaple–Bideford area was marked by the '*Goniatites spiralis*' Bed, well exposed at Fremington and indicative of the P_{2a} Zone (upper Viséan). The bed yields

(A 10546)

A. Tintagel Castle, Cornwall

Plate 3

(*For full explanation see p. ix*)

B. Inverted Lower Carboniferous limestone, Meldon Pool, Okehampton

(A 10061)

A. Lower Carboniferous ridge looking west towards Drewsteignton

Plate 4   (*For full explanation see p. ix*)

B. Zig-zag folds in Crackington Formation, Millook Haven

Posidonia corrugata, Caneyella membranacea, Neoglyphioceras spirale and *Mesoglyphioceras granosum.*

Viséan strata of north-east Devon have been divided by J. M. Thomas (1963) into two facies of roughly equivalent age. The Bampton Limestone Group comprises cherts with interbedded limestones and shows some evidence of deposition by turbidity currents; the Westleigh Limestones comprise shales with thickly bedded turbidites which consist not of quartz grains but of tiny fragments of shells and limestone cemented by calcite and may be termed limestone sandstones or calcarenites. The beds at Bampton comprise lower and upper limestones separated by radiolarian slaty cherts. Shales associated with the upper limestones show the '*spiralis*' bed and have yielded a trilobite fauna including a species of *Phillipsia*. The highest Viséan fossils recorded from the Bampton area are spirally ornamented goniatites similar to *Mesoglyphioceras granosum* which is referable to the P_{2a} Zone. Quarries near Westleigh show thickly bedded upper calcarenites overlying fine-grained lower calcarenites, but to the north the upper group appears to rest on black slates. Shales associated with the limestones show worm burrows and have yielded many goniatites, orthocones and *Posidonia* and a few productoids, chonetoids and trilobites. The goniatites of the Westleigh Limestones range from II to P_{2a} age.

No trace of the volcanic beds so prominent along the southern edge of the great Culm basin has been found in north Devon.

Upper Carboniferous
Crackington Formation

The term Crackington Formation was adopted to include Ussher's Exeter-type Culm, a sequence of shales with thin turbidite sandstones apparently approximately corresponding to the Namurian Series (Plate 4B). They were laid down in a basin on whose floor unconsolidated sediments were period-ically disturbed by currents of turbid muddy water. Such turbidity currents swept down the flanks of the basin carrying sands and even pebbles; they com-monly originated as an influx of sediment-laden flood water, but may also have been caused by slight earth movements disturbing unstable deposits on high submarine slopes. Flow occurred along the sea floor and gradually ceased as, with deposition of its load, the density of the current approached that of the surrounding water. The scouring action of such currents, and their deposition of higher density sediment on soft muds, produced a variety of sedimentary structures from which it is possible to deduce whether beds are right way up or inverted and, less commonly, the direction of flow of the turbidity currents. Evidence from the area north and north-west of Oke-hampton indicates that the main direction of flow was westwards along the axis of the geosyncline, although currents flowed in other directions from time to time. Turbidity currents are commonly thought to occur in deposi-tional basins at least 600 ft deep, but there seems no reason why they should not have flowed beneath shallower water, when they may have been slower and less dense.

Despite the new fossil localities discovered by the Geological Survey in recent years the fauna of the Crackington Formation is still not well known. Fossils found so far on the south side of the synclinorium span almost the whole of the Namurian and part of the lower Westphalian. Between Boscastle and Bude the turbidite sequence (Crackington Measures of Ashwin 1958) contains goniatites ranging in age from the E_{2c} to R_{2c} zones, and in the Oke-hampton district from E_2 to R_{2b}; around Exeter, Butcher and Hodson (1960) have identified goniatites of H_1 to R_{2b} ages from sections at Idestone Hill, Bonhay Road, Stoke Road, Pinhoe brickpit and Perridge Tunnel. There is so far no record of goniatites from the youngest or oldest Namurian stages in the southern part of the Carboniferous outcrop. However, shales and thin sandstones of the Crackington Formation at Wanson Mouth, near Bude, have yielded goniatites of the *Gastrioceras listeri* group and hence the top of the formation here lies in the Westphalian.

Ussher mapped his Eggesford-type Culm, consisting of evenly bedded sandstones and shales like those of the Crackington Formation in the south, between Eggesford and the coast north of Bude. It underlies land more poorly drained and less easily worked than that on the Bude Formation. Generally, such fossils as have been found in the Eggesford-type Culm are about the same (basal Westphalian) age as those from the Crackington Formation at Wanson Mouth, as are those from the Welcombe Measures of Ashwin (1958) on the coast south of Hartland Point.

Shales near Thornham, about midway between Chulmleigh and South Molton, have yielded specimens of *Caneyella* and *G.* cf. *listeri*; a fauna from Hescott Quarry, two miles west of Clovelly, belongs to the *G. subcrenatum* horizon at the Namurian–Westphalian boundary; and goniatites of the *Gastrioceras circumnodosum* group from near Knap Head, Welcombe (Butcher and Hodson 1960), suggest a G_2 age. The presence in the Hartland–Clovelly area of the marine bands of '*Gastrioceras*' *sigma* (*Donetzoceras sigma*), *G. cancellatum* and *G. cumbriense*, in addition to those of *G. subcrenatum* and *G. listeri*, is suggested by Freshney and Taylor (1973). On lithological and faunal grounds the Eggesford-type Culm and the Welcombe Formation are now referred to the Crackington Formation.

In the Bampton area specimens of *Cravenoceratoides* of E_2 age were found in black shales above the Bampton Limestones (House and Selwood 1966) and south of Bampton *Reticuloceras superbilingue* (R_{2c} age) has been recorded. In north Devon the Lower Carboniferous cherts and limestones are succeeded by a shale and siltstone sequence (Limekiln Beds of Prentice 1960a), which at Fremington has yielded specimens of *Reticuloceras* of the R_1 and R_2 stages (Moore 1929; Prentice 1960a). The Instow Beds, of greywacke and shale, overlie the Limekiln Beds in the Barnstaple–Bideford area (Prentice 1960a and b); their lower part may be of Namurian age but the main fossil evidence comes from a fish bed near the top of the sequence which has yielded *Gastrioceras circumnodosum* and was equated by Prentice (1960a) with the *G. listeri* horizon of the lower Westphalian. The Fish Bed has also yielded *Anthracoceratites arcuatilobum*, orthocones, *Dunbarella* and the fishes *Rhabdoderma elegans* and *Elonichthys aitkeni*. Other faunas of lower Westphalian age have been collected between Hartland Point and Westward Ho!

According to Prentice, the Instow Beds are succeeded to the south by the Northam Beds, a Coal Measures facies of shales, siltstones and sandstones (Westphalian age), but work by De Raaf, Reading and Walker (1965) and Reading (1965) led them to suggest that their Westward Ho! Formation, of

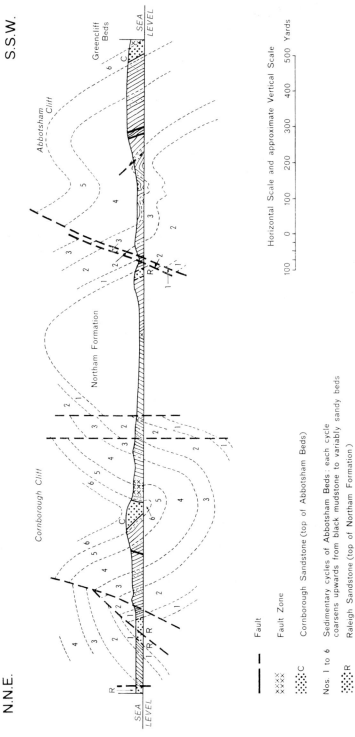

Fig. 13. *Cliff section south-south-west of Westward Ho!*

After J. F. M. De Raaf, H. G. Reading and R. G. Walker 1965

silty mudstones and turbidites, and the overlying Northam Formation are together equivalent to Prentice's Northam Beds. However, there are no mappable distinctions between Limekiln Beds, Instow Beds and Westward Ho! Formation and it seems best to include them all in the Crackington Formation.

Bude Formation

The Bude Formation is equivalent to Owen's (1934) Bude Sandstones and to Ussher's Morchard-type Culm. It comprises thickly bedded and massive sandstones with siltstones and shales. Few fossils have been recorded from these beds. A fauna from two localities near Morchard Bishop and Eggesford includes *Caneyella sp.*, *Dunbarella sp.* and *Anthracoceratoides cornubiensis* and has been correlated (Ramsbottom 1965, *Summ. Prog. Geol. Surv. for 1964*) with that of the Margam Marine Band, which is of lower Westphalian age and in South Wales lies just above the horizon of *Gastrioceras listeri*. A similar anthracoceratid has been found at Sandy Mouth, two miles north of Bude. Freshney and Taylor (1973) have established a Bude Formation succession containing several nodular shales which may, from the thicknesses of strata present and comparison with the South Wales sequence, represent marine bands above Westphalian A. The formation has also yielded Westphalian plants (Arber 1907; Crookall 1930) and contains a fish bed at Bude (Owen 1951).

Prentice (1960b) suggested that in north Devon his Northam Beds were succeeded to the south by sandstones, siltstones and shales of the Abbotsham Beds, which contain fish and plant remains. Non-marine bivalves from the Zone of *Anthraconaia lenisulcata* have been collected from these beds in Abbotsham Cliff, and others from the Zone of *Carbonicola communis* from Robert's Quarry at Bideford. The *G. amaliae* Marine Band (Ramsbottom and Calver 1962) has been identified from cycle 3 of the Abbotsham Beds, from Westacott Cliff, as well as from between Hartland Quay and Hartland Point. The Bideford Group of De Raaf, Reading and Walker (1965), comprising the Northam and Abbotsham formations, when traced inland towards South Molton, passes imperceptibly from the paralic facies of the coast into typical Bude Formation deposits, but the Bideford Group lithologies are sufficiently distinct to merit a separate name—Bideford Formation. A considerable thickness of Bideford Formation rocks occurs below the horizon of *G. amaliae*, and this probably represents a thickening of the sequence, since no occurrence of the *G. listeri* horizon, a persistent marker band in the area, has been found. To the south of, and possibly overlying, strata of the Bideford Formation are Prentice's (1960b) Greencliff Beds, now included in the Bude Formation. His Cockington Beds contain strata now classified partly as Crackington Formation and partly as Bude Formation, and both *G. listeri* and *G. amaliae* have been found in them.

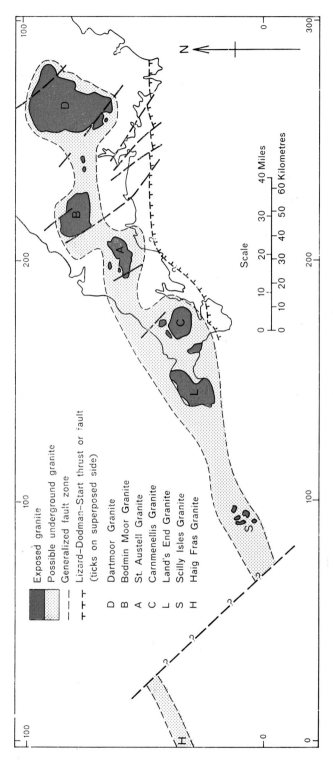

FIG. 14. *Possible underground connections of the Armorican granites* Based mainly on M. H. P. Bott, A. A. Day and D. Masson Smith 1958

development of recent years is the recognition of the growth in the solid rock of secondary potassium feldspar replacing plagioclase, a feature characteristic of the granites of Devon and Cornwall. The big-feldspar granite of Dartmoor shows finer grained variants, particularly near the margins and roof and perhaps resulting from contamination of the magma, and local fine-grained chilled contacts, as seen in the East Okement River. A general division into coarsely megacrystic and finer poorly megacrystic rock has also been recognized in the Cornish granites by various geologists, who have generally considered the finer grained facies to be the younger. Probably all have similar histories to that of Dartmoor.

The probable underground shape of the Dartmoor mass suggests that the present high tors lay near the top of the original laccolith, and in Cornwall roof rocks have been preserved at Porthmeor, Rinsey and Carpalla. The tabular joints typical of most tors may parallel the roof of the intrusion or flow surfaces or stress surfaces in the consolidating magma. Although they commonly appear to follow the topography it is unlikely that they are a product of weathering.

Earlier editions of this book suggested that the original positions of the granite bosses were governed by existing structures, but the tectonics of the country rocks are so complex that this possibility is not demonstrable.

Minor Acid Intrusions

All the granites contain felsic pods and veins. Segregations of quartz-tourmaline rock are common and dykes of this material traverse the country rock (p. 48). Fine-grained aplitic rocks occur as apparent dykes and sills ranging from a few inches to 20 ft or so thick. They comprise potash and plagioclase feldspar, and more quartz and less biotite than occur in the main granites; megacrysts are few but may lie across contacts. The origin of many, particularly the smaller occurrences, appears to be magmatic, and Brammall and Harwood considered these rocks to belong to a final intrusive phase. However, some of the masses are of irregular shape, show no effects of marginal chilling and may even grade into surrounding granite, and contain plagioclase feldspar of a type which is found in the main granites and which would be unlikely to crystallize from late residual fluids of the magma. Perhaps a few, at least, of these bodies represent country-rock sandstones caught up in the advancing magma.

Sodium-rich aplite dykes and veins are associated with the Dartmoor Granite. They are pale grey and comprise albite, quartz and orthoclase-perthite with micas and accessory topaz and tourmaline. The mica may be muscovite or the lithium-bearing variant lepidolite. Generally the dykes are very small but the Meldon Aplite, which intrudes Lower Carboniferous rocks on the north-west flank of Dartmoor, is up to 60 ft thick and contains an exciting range of accessory minerals, including petalite, topaz, fluorite, bavenite, pollucite, spodumene and several beryllium minerals. Lepidolite from this aplite has been found by potassium-argon determinations to be 254 ± 5 million years old (Miller and Mohr 1964). However, leakage of argon from the mica is a possible source of error and application of this method has been found to yield lower ages for the muscovite of south-west England granites than for the biotite; for example, the figures for the granite

(*A 10366*)

A. Black Rock, Widemouth Bay

Plate 5

(*For full explanation see p. x*)

B. Dolerite dyke intrusive into Lower Carboniferous rocks,
Meldon Quarry, Okehampton

(*A 10062*)

(A 10432)

A. Black-a-ven Brook, Dartmoor

Plate 6 (*For full explanation see p. x*)

B. Blackingstone Rock, a tor of Dartmoor Granite

(A 10059)

of Kit Hill are 265 ± 6 and 296 ± 8 million years, and consequently the Meldon Aplite has not been proved to be appreciably younger than the Dartmoor Granite.

Pegmatites have developed locally within the granites, and in places occur associated with aplites, as at Tremearne, although they are less common than the latter. Their large crystals of feldspar, quartz, mica and tourmaline, which locally have grown in herring-bone or chevron patterns, make them attractive rocks but very few have been worked for their minerals. The tourmaline pegmatite at Knill's Monument, near St. Ives, and the locally coarsely crystalline greisen of Cligga Head are among the best developed. Formerly a pegmatite with biotite crystals measuring up to six inches in length was worked for its lithia content on Trelavour Downs, near St. Dennis; another, 50 yd wide on Tresayes Downs, near Roche, and consisting of alkali feldspar with a little quartz and tourmaline, was worked for the manufacture of glass. Topaz and apatite are characteristic of some pegmatite dykes, and the clear blue topaz of the greisens at Cligga and St. Michael's Mount is well known. An interesting example showing a relationship between the pegmatite dykes and the mineralized veins was brought to light in the wolfram openworks at the northern end of Bodmin Moor where large crystalline masses of wolfram formed a constituent of orthoclase-quartz-pegmatites. Unusual pegmatites occur near Ponsworthy (red orthoclase and green chloritized feldspar with quartz and tourmaline), on Bittleford Down (feldspar, quartz, hornblende and sphene) and at Tremearne (large crystals of feldspar and tourmaline perpendicular to the surface of the pegmatite).

Elvans, a term best restricted to quartz-porphyry dykes or more rarely sills, are of general occurrence around the granites; possibly, they represent feeders for subaerial vulcanicity of dacite-rhyolite type. In addition to quartz the megacrysts may include feldspar and rarely mica; they lie in a fine-grained rhyolitic base. The dykes have chilled margins, range up to 150 ft in thickness and in places form steep-sided ridges rising boldly from the lower ground of the sediments as at Coomb, near St. Kew. Among the most handsome of these rocks are the Tremore elvan, which is spangled with purple fluorspar, the pinitiferous elvan at Goldsithney, and the Praa Sands elvan. A flow-banded quartz-porphyry occurs at Tregonetha. Rocks of this type without the megacrysts are termed felsites and they may in general be earlier than the porphyries.

Lamprophyres

Lamprophyre dykes extend farther from the main granites than do the minor acid intrusions. They differ from the elvans and felsites mainly in containing more mafic grains, commonly biotite but locally olivine or other ferromagnesian minerals. Most of the dykes are thoroughly decomposed and comprise quartz grains and possibly a few quartz xenocrysts with clay minerals, chlorite and iron oxides. Such rocks are clearly exposed in Newquay Headland and the Gannel, while others, as at Hicksmill and Lemail, are fairly fresh biotite-orthoclase traps. A weathered dyke trends east-north-east through Halwill for over four miles, and several shorter ones occur thereabouts and between Hatherleigh–Northlew and Exeter. They cut across the strike of the enclosing Carboniferous rocks and are thought to occupy feeder channels for the Permian lavas. The rock at Hicksmill—brownish pink, glistening with large crystals of biotite and speckled with small pink orthoclase—is perhaps the most beautiful lamprophyre in the west country.

Xenoliths

Only the granites contain inclusions of country rock of appreciable size, most of which are darker than the granite and ovoid in shape and show the effects of potash metasomatism. Some may originally have been basic igneous rocks and appear as biotite-rich masses with much plagioclase feldspar, less quartz and orthoclase than the granite, and a good deal of zircon and apatite. Inclusions of shale have been converted to fine-grained quartz-feldspar-biotite rock with cordierite, andalusite, sillimanite, spinel and corundum. Sandstone masses engulfed by the magma would be expected to change into xenoliths of aplitic material, which suggests that some of the fine-grained granitic rocks long assumed to be truly igneous may be of sedimentary origin (p. 46). Outcrops and boulders on Throwleigh Common, northern Dartmoor, and near Harford on southern Dartmoor, suggest the presence of rafts up to two miles long, of fine-grained 'granite' with megacrysts of quartz and feldspar, which are locally called 'elvan' but may be gigantic xenoliths.

Alteration of the Granite

During its final stages of cooling at depth the granite gave off active solutions and volatile gases above their 'critical temperatures', including great volumes of fluorine and boron along with sulphur dioxide, super-heated steam and carbon dioxide. These emissions entered fissures in the consolidated granite and the adjacent rocks to give rise alike to the mineral lodes and to three main types of alteration, tourmalinization, greisening and kaolinization, which overlap and are not always distinct from one another.

Boric vapours attacked the calcareous muds and limestones adjacent to the granite and partly converted them into axinite and garnet rocks, as at Tremore; where lime was absent a quartz-schorl hornfels was formed that retained the structures of the sediment. Tourmaline is rich in boron, containing as it does up to 10 per cent of boric acid. It commonly grew in the granites in primary and secondary generations; the first is represented by irregular grains of various sizes and by tiny inclusions in biotite, and the second by crystal needles related to joints and veins. In the process the original feldspars and micas suffered; where mica was replaced a tourmaline-granite was formed, but frequently the process continued and the feldspars also were replaced so that a quartz-schorl rock was all that remained of the original granite. The most outstanding instance of tourmalinization is Roche Rock (Plate 7A), but similar dykes form bold features as at Devil's Jump, near Camelford, and Lanlavery Rocks, north of Bodmin Moor. Varying degrees of tourmalinization of granite can be seen, as in 'luxullianite', 'trevalganite' and 'trowlesworthite', the last rock also containing fluorspar. Veins of tourmaline and quartz cut the kaolinized granite in all directions and offer hindrances to the china clay workers, by whom they are called among other terms, 'stent'. Their mode of occurrence as veins traversing the kaolinized rock indicates an origin earlier than the kaolinization.

Luxullianite was originally a big-feldspar granite, and the megacrysts of pink feldspar stand out in striking contrast to the dark blue groundmass of quartz and schorl. Its value as a decorative stone may be gauged by the tomb of the Duke of Wellington in St. Paul's Cathedral.

Greisening too occurred in two phases. In the first the feldspars were replaced by commonly pseudomorphous aggregates of quartz and white mica, usually a lithium-bearing variety. Fluorine-bearing solutions or vapours produced the second phase which is associated with joints and lodes and has produced rocks of quartz and white mica with topaz and fluorspar and rare tourmaline. Some greisen veins in the granite have been worked for tin, as at Beam and Bunny near St. Austell. The process of greisening appears to have operated earlier than that of kaolinization (Plate 11B).

All the granite masses are in part kaolinized but that of St. Austell (Fig. 15) has suffered most. The process, involving change into china clay, was probably the last of the hydrothermal alterations and was largely effected by the movement of acid solutions along joints; near St. Austell these fissures form a reticulate pattern and have been much excavated. The crystals of plagioclase feldspar have undergone the greatest change and under the microscope they can be seen to consist of aggregates of minute scales of kaolinite with some secondary feldspar and quartz. The potash feldspar megacrysts were more resistant than the matrix to this change and many have remained unchanged or only partly altered in a rock that crumbles down in the hand to powder. Strangely enough, however, locally they were completely converted into masses of sericitic mica retaining the original crystalline form of the feldspar. Such specimens are normally twinned and are called by the clay workers 'pigs' eggs'.

Some parts of the St. Austell Granite, notably between Treviscoe and Nanpean, have undergone changes which have produced china stone, apparently by the action of fluorine and lithium upon an albite-rich granite. China stones are cream coloured and commonly tinged purple by fluorspar. They are classed according to their hardness as 'hard purple' and 'mild purple', whilst those devoid of fluorspar are grouped as 'hard white' and 'mild white'. A weathered material called 'buffstone' was the rock first worked. These rocks consist of white mica, in both primary large plates and small secondary scales, of the varieties lepidolite and gilbertite. Primary albite forms small idiomorphic crystals commonly enclosed in quartz, microperthite is abundant and topaz occurs as large primary grains. In some varieties purple fluorspar forms much of the rock and spreads along the cleavages of muscovite plates.

Contact Metamorphism

The Armorican granite bosses baked country rocks up to 4 miles from contacts, the effects varying with the nature of the host rocks. Thus argillaceous Devonian strata around the Land's End Granite were converted to quartz-cordierite-biotite-hornfelses which bear minor sapphire. Andalusite commonly takes the place of cordierite, as in the Carboniferous shales skirting the north of Dartmoor. In the outer parts of aureoles, spotting in argillaceous strata may be the only clear evidence of metamorphism. Small

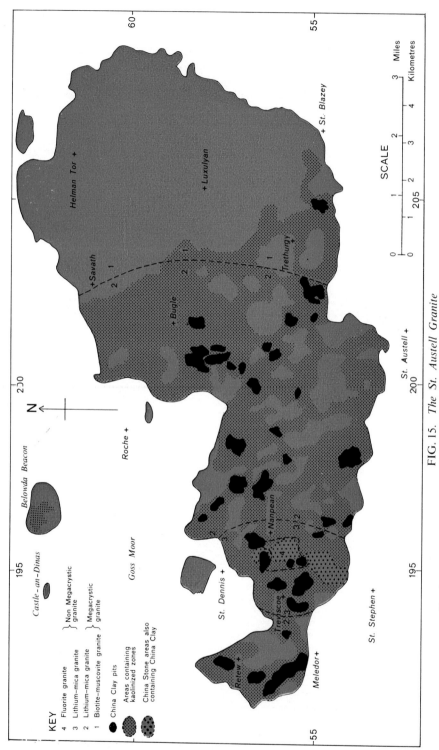

FIG. 15. *The St. Austell Granite*

Granite types after C. S. Exley 1959. China Clay and China Stone details by C. M. Bristow

amounts of andalusite or cordierite occur in the more argillaceous sandstones, but localized growth of muscovite or biotite constituted the main mineralogical change in these rocks. Minerals like amphibole, pyroxene and garnet developed in calcareous rocks, which gave rise, for example, to the hard, whitish grey, quartz-rich calc-flintas in the aureoles of the St. Austell, Bodmin and Dartmoor granites. The only limestones to have undergone contact metamorphism are those of the Lower Carboniferous around the northern margin of Dartmoor. They have been recrystallized and, where associated with chert and calcareous shales, have given rise to pyroxene-, garnet-, wollastonite-, and calcite-bearing calc-silicate hornfelses. Minerals present in small amounts include albite, scapolite, idocrase, bustamite and rhodonite. Boron from the magma has facilitated the development of tourmaline within the aureoles, and both arenaceous and argillaceous rocks have been converted to banded quartz-tourmaline-hornfels that may superficially resemble schorl-rock. Axinite rather than tourmaline has formed in calcareous beds.

Contact metamorphism of greenstones has generally produced hornblende-plagioclase rocks, the hornblende subsequently being replaced by biotite where potassium was available from the magma. The striking cordierite-cummingtonite-anthophyllite-biotite-hornfelses of Kenidjack and Botallack, Land's End, represent a departure from the metamorphic trend evident in the majority of greenstones. Similar hornfelses occur in the north-western sector of the Dartmoor aureole, but here they have formed from silty, argillaceous tuffs of keratophyric affinity. Other metamorphosed greenstones bear pyroxene, garnet, and epidote, as at Botallack and Tater-du, Land's End. The presence of tourmaline and axinite in igneous hornfelses is indicative of boron metasomatism.

Both igneous and sedimentary rocks were locally affected by late-stage hydrothermal activity, which produced mainly chlorite, quartz, and sericite or muscovite.

6. New Red Rocks

The controversy over the division of the generally unfossiliferous, red-coloured breccias, conglomerates, sandstones and marls which crop out over much of east Devon and west Somerset into the Permian and Triassic systems has perpetuated the use of the title 'New Red Sandstone Series' under which De la Beche and Godwin-Austen earlier described these rocks. Their outcrop, including the larger outliers at Slapton, Hollacombe and Portledge, is shown in Fig. 1. Several smaller outliers of Permian rocks occur between Bolt Head and Plymouth and off-shore studies indicate that the New Red Rocks must formerly have covered a considerable area to the south; a thickness of some 3100 ft of these sediments is believed to underlie the English Channel south-east of Looe, and similar rocks are reported north-west of Bideford.

In contrast to the 'dunland' areas of central Devon the characteristic red soils derived from these rocks lend warmth and charm to the local scenery (Plate 7B) and, on the south Devon coast, form a great part of the colourful and locally grand cliffs between Paignton and Seaton; over 9000 ft of beds are exposed along these sections.

Permian

Following Conybeare and Buckland's comparison of the Heavitree Conglomerate (near Exeter) with the Rothliegende formations of the Hunsrück district many of the early geologists, including Murchison, Hull and Irving, were in favour of correlating the breccias of Devon with the Lower Permian of Germany. Ussher (1900; 1906) suggested a possible correlation which also utilized the contemporaneous basic lava flows, generally referred to as the Exeter Volcanic Series or Exeter Traps. Though common to both Germany and Devon the vulcanicity was far more intense in the former. Sherlock (1947) was sceptical of using the lavas as evidence of age and earlier, in attempting to disestablish the use of the term 'Permian', had pointed out that the Rothliegende pass down into the Carboniferous. Isotopic dating of lavas from Killerton and Dunchideock—two members of the Exeter Volcanic Series—indicates that the volcanic activity took place at or near the Permo-Carboniferous time boundary. There is, therefore, a strong probability that parts of the lower breccia sequences in both south and central Devon are of Stephanian (Upper Carboniferous) age.

Great differences of opinion developed among the earlier writers as to the upward limit of the Permian in Devon. Irving (1888) argued that the junction occurred at the base of the Budleigh Salterton Pebble Beds, a horizon which has been taken as the equivalent of the Bunter Pebble Beds of the Midlands; Ussher, in accepting this contention, suggested that the break beneath the Pebble Beds could equally well represent that between Bunter and Keuper. However, no more precise limit has been defined. The succession of the Permian rocks of the south Devon coast is as follows:

5. Red Marls
4. Dawlish Breccias and Lower Sandstones
3. Teignmouth Breccia
2. Watcombe Conglomerate
1. Watcombe Clay.

Laming (1966) has proposed group and local names for these beds.

Following the deformation of the Devonian and Carboniferous rocks and the emplacement of the Dartmoor Granite by early Stephanian times (about 290 million years ago) the region underwent severe denudation, probably under hot desert conditions. Angular fragments and coarse sand were swept down from mountains in the north, west and south by torrential streams to accumulate as thick fans at the mouths of the valleys; finer detritus was probably transported beyond the fans and laid down in broad depressions and temporary lakes such as that which underlay the present Crediton valley. The mainly southerly derivation apparent in south Devon was confirmed recently when Laming (1965; 1966) indicated that the breccias overlap progressively northwards. Scrivenor (1948), however, suggested the rocks were beach deposits laid down by a sea which advanced westwards and eventually encroached on a more arid regime north of Exeter. In the Crediton valley material was transported mainly from the Dartmoor and Exmoor areas.

Resting unconformably on the older rocks, and filling the relict topography, many of the basal deposits contain fragments mainly of local origin. Successive horizons of the breccio-conglomerates contain pebbles and mineral assemblages directly related to the gradual denudation of the granite's mantle, though fragments of the granite itself are rare. It is thought that the unroofing of the granite was not completed until Wealden times but the presence of kaolinite, in both Watcombe Clay and some of the finer sediments of the Crediton valley, suggests that some surface manifestations of the granite were in fact being denuded in late Stephanian and early Permian times.

Watcombe Clay

The Watcombe Clay seems to be the earliest New Red deposit exposed at the surface in south Devon and occupies small areas near Watcombe, Barton and Daccombe between Teignmouth and Bishopsteignton. On the coast the dark red silty clays are seen to be overlain by breccio-conglomerates amongst which they are brought up by faults. The clays have long been used locally in making terra-cotta ware, tiles and bricks.

Inland these clays lose their distinctive characters, becoming a loamy or clayey breccia similar to the Cadbury Breccia of the Crediton district.

Watcombe Conglomerate and Teignmouth Breccia

Ussher described the succeeding breccio-conglomerates and boulder breccias under the names Watcombe Conglomerate and Teignmouth Breccia respectively. The lower group is characterized by the abundance of fossiliferous Devonian limestone fragments, and lesser amounts of sandstone, weathered igneous rocks and vein quartz; these beds form the cliffs between Babbacombe and the Teign estuary and are well exposed at Watcombe and Petit Tor Crags. Thick sandstones occur within the group, as at Oddicombe and around The Ness. The upper beds are distinctive in that large boulders of a peculiar red quartz-porphyry are by far the commonest detrital constituents—a rock type thought by Ussher to occur near Christow; these boulders are set in reddish brown earthy sand and fine gravel. Long lists of the other rock types present in these breccias have been published. The cliffs between Teignmouth and Holcombe Tunnel are formed of these beds which, in a much-faulted section between the tunnel and Horse Cove, are brought into contact with various horizons of the overlying Dawlish Breccias.

Dawlish Breccias and Lower Sandstones

These beds crop out between Dawlish and Exmouth as a series of irregularly intercalated fine breccias and red sandstones in which many flesh-coloured cleavage fragments of sanidine ('murchisonite' of early writers) are found. To the north, around Exeter, the uppermost breccia horizon (Heavitree Conglomerate) is succeeded by a series of sandstones. Breccias bearing sanidine fragments can be traced from the coast into the Crediton valley where they form the uppermost beds of a litho-facies succession divided into four by Hutchins, on the basis of the pebble and heavy mineral assemblages of the breccias. His classification is given below, to the right of the modified one adopted by the Geological Survey.

Crediton Conglomerates {	St. Cyres Beds Crediton Beds
Knowle Sandstones } Bow Conglomerates }	Bow Beds
Cadbury Breccia	Cadbury Breccia

The Crediton trough, the larger of the two westward extensions of the main Permian outcrop, probably developed along a major basement structure extending to the Atlantic coast. The sediments and lavas were laid down in a broad depression which subsequently was trough-faulted west of Bow; west of the River Okement the Permian rocks are separated into a series of outliers by later north-west-trending faults. A maximum depth to the trough of 1500 ft has been indicated at North Tawton by geophysical methods.

Cadbury Breccia

Loose basal gravels and clayey breccias (Cadbury Breccia) are not well developed west of Sandford, but north and west of Silverton these deposits contain fragments of Pilton Beds derived from the north-west, together with locally-derived Upper Carboniferous debris.

Bow Conglomerates and Knowle Sandstones

West of Bow the Crediton trough is filled with breccio-conglomerates of Bow Conglomerates age. Lamprophyric and basaltic lavas occur near the margins and within a group of red sandstones (Knowle Sandstones) east of the village; these sandstones are believed to be partly equivalent to and

(A 407)

A. Roche Rock, Roche, a quartz-schorl dyke

Plate 7 *(For full explanation see p. x)*

B. Permian dip and scarp scenery, Colebrooke

(A 10108)

Plate 8

A. Permian volcanic neck, Hannaborough Quarry (*A 10094*)

(*For full explanation see p. x*)

B. Pebble Beds and Triassic sandstone, Budleigh Salterton (*A 10113*)

partly younger than the breccio-conglomerates. The lavas are members of the Exeter Volcanic Series (see below) and, if correlated with those of Dunchideock and Ide (which Ussher showed to lie at the base of the Watcombe Conglomerate) and the contemporaneous vulcanicity at the same horizon near Torquay, overlie a great thickness of Upper Carboniferous red beds in the Crediton valley, whilst the Watcombe Clay represents sedimentation of the same age to the south.

Givetian (Middle Devonian) limestone pebbles occur in coarse conglomerates of a supposedly low Bow Conglomerates horizon at Solland and Westacott quarries, west and east of North Tawton; the source of the limestone is not known.

Crediton Conglomerates

Apart from the absence of sanidine fragments Hutchins's Crediton Beds are indistinguishable from his St. Cyres Beds and have not been separated by the Geological Survey on their maps. These breccio-conglomerates contain fragments of tourmalinized slate and lava, untourmalinized lava and locally-derived Carboniferous sediments.

Red Marls

South-east of Exeter a series of red marls with thick red and whitish grey sandstones overlies the Dawlish Breccias, though the exact nature of the contact is uncertain. Between Exmouth and Straight Point these beds are well exposed in a much-faulted cliff section; at Straight Point a fault brings in a series of marls, without sandstones, at least 500 ft thick, which in turn is overlain near Budleigh Salterton by the Triassic Pebble Beds without any apparent disconformity other than slight irregularities in the junction consistent with the change in the sedimentological conditions.

Exeter Volcanic Series

The several types of volcanic rock grouped together under this title crop out as isolated remnants (Fig. 16) of formerly extensive lava flows extruded at and soon after the start of Permian times. Small areas of igneous rock near Plymouth and Kingsbridge may also be of Permian age. The series occurs as a suite of vesicular and scoriaceous lavas, predominantly potash-rich but with a subordinate group of olivine-basalts, together with small outcrops of vent agglomerate (Plate 8A). Originally described by Ussher and Teall, and later by Hobson, Hatch and Tidmarsh, these 'trappean' rocks have been subject to a recent re-appraisal by Dr. Diane C. Knill. In general they have been extensively carbonated and hematitized and although the original orthoclase and plagioclase feldspars retain their characters the ferromagnesian minerals (biotite, augite and olivine) are almost completely decomposed. Biotite may rarely be unaltered; olivine on the other hand is typically represented by a red mass of iron oxide and carbonate called by Tidmarsh 'iddingsite'. The zeolites analcime and natrolite occur in some of the lavas; in others amygdales are filled with montmorillonoid clay minerals or occasionally illite.

Tidmarsh (1932) divided the volcanic rocks into the following series:

(i) Basic, intermediate and lamprophyric lavas with xenocrysts of quartz, feldspar, pyroxene, biotite, 'iddingsite' and apatite (Hatherleigh Series).

(ii) Basic, intermediate and lamprophyric lavas containing xenocrysts of biotite and 'iddingsite' (Pocombe Series).

(iii) An unrelated group of potassic, basic minettes with various xenocrystal assemblages.

These he regarded as "hybrid types resulting from the admixture of a late acid residuum and the [basic] depth-residuum of the Dartmoor igneous mass, both fractions containing at the time of admixture solid as well as liquid phases". His classification is unnecessarily complex since it cuts across the natural grouping into which these lavas fall and also requires an abnormal crystallization sequence of the parent magma. Dr. Knill, who divides the lavas into (a) Basaltic Group (olivine-basalts and dolerites) and (b) Potash-rich Group (trachybasalts, syenitic lamprophyres, minettes and leucite-bearing rocks), tentatively invokes an olivine-basalt body at depth, subsequently differentiated to produce an olivine-augite-peridotite and contaminated by the assimilation of potassic granite, to account for the genesis of the potassic lava suite. Hawkes (*in* the Okehampton Sheet Memoir), in relating the sequence of igneous events in the north Dartmoor district to the region as a whole, has suggested that supposed basic material at shallow depth around the northern margin of the granite is the result of sub-crustal melting concomitant with the rise of the granitic material, and that its emplacement roughly coincided with the intrusion of the granite batholith itself. Basic magmas rising during this period may have been the major source for the basal Permian extrusives, but since the earliest lava flows, north of Okehampton, are potassium-rich lamprophyres the parent magma must have been modified before extrusion took place. Tidmarsh suggested that the modification resulted from the assimilation of either granite magma or crystallized portions of the Dartmoor mass at depth, a concept supported by the abundance of quartz xenocrysts in the lamprophyres and the presence of granite xenoliths in the minette lavas at Killerton Park. However, to account for the high potassium content (up to 13 per cent K_2O) of the potassic types, particularly the minettes, it is suggested that the parent basic magma was contaminated at depth by potassium-rich aqueous fluid escaping from the Dartmoor magma. This then would have produced an extrusive magma of minette composition; subsequently a decrease in interaction between the supposed basic magma and the hydrous fluids could have given rise to the later, more basic, olivine-basalt flows.

Most exposures in the igneous rocks are poor. Amongst the better ones, where the relationship to the Permian sediments can be seen, are quarries at Stone (North Tawton) and Webberton Cross, both in porphyritic olivine-basalts; Hannaborough Quarry, near Hatherleigh, showing olivine-biotite-minette and vent agglomerate; the outcrops of trachybasalts at Uton, Posbury Clump, and Crossmead (Pocombe); several localities at Killerton Park, where augite-biotite-minette and biotite-apatite-minette occur; and others around Columbjohn and Budlake showing the various syenitic lamprophyres.

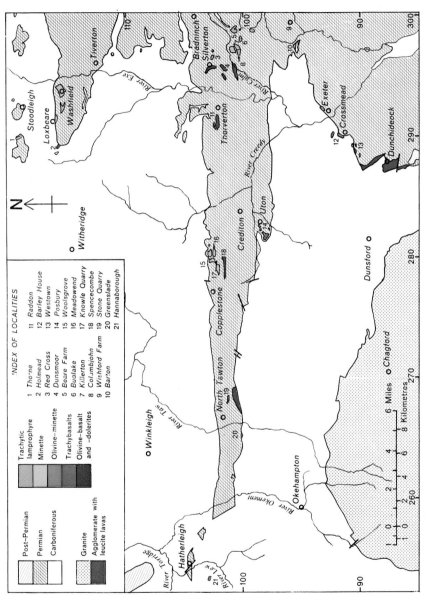

FIG. 16. Distribution of the Exeter Volcanic Series in mid-Devon

Ophitic olivine-dolerites, the second member of Knill's 'Basaltic Group', crop out mainly around Silverton and Thorverton and are best exposed at Raddon Quarry. In the Crediton valley similar rocks crop out along a low ridge between Knowle village and Spencecombe; immediately to the north of this ridge minettes occur which are essentially of an augite-biotite-olivine type and are exposed at Knowle Hill (Crediton), Meadowend and Woolsgrove. Knill has described rocks from these localities as "mafic syenites of yogoitic affinity" and believes that the minettes of the Washfield–Loxbeare–Thorne district north-west of Tiverton are transitional types between those at Knowle Hill and Killerton.

An agglomerate near Holmead Farm, 5 miles north-west of Tiverton Station, has yielded, in addition to fragments of minette, pieces of vesicular basaltic leucite-bearing lavas akin to orendite and olivine-leucitite.

Owing to extensive alteration many of the lavas appear similar in the field; nearly all are vesicular but they range in colour from dark bluish grey, grey and purple to red and reddish brown. Trachytoid texture is preserved in some types of both lamprophyre and basalt, while veins of dolomite are not uncommon.

Triassic

Although the Lower Mottled Sandstones of the Midlands Bunter succession do not appear to be represented anywhere in the region, the lowest Triassic beds rest upon Permian marls without any sign of unconformity. In the Vale of Porlock and the Watchet districts of west Somerset successive horizons of the Triassic have been shown to overlap on to a denuded landscape of Devonian rocks. In some instances where the marginal facies rests on Carboniferous Limestone a basal conglomerate, known as the 'Dolomitic Conglomerate', is formed.

Between Budleigh Salterton and Culverhole Point, east of the Axe estuary, representatives of the Bunter and Keuper formations are admirably exposed in almost continuous cliff sections.

Pebble Beds

The Pebble Beds (Plate 8B) exposed at Budleigh Salterton consist of a false-bedded sequence 70 to 80 ft thick; the strata are made up of pebbles and scattered subangular fragments of quartzite, grit and quartz intercalated with dark red and sometimes grey and buff sands. Some of the quartzite pebbles have yielded the fossils *Lingula lesueuri* and *Orthis budleighensis*, species respectively typical of the Grès Armoricain and Grès de May (Ordovician) of north-west France. Pebbles of Devonian rocks containing *Cyrtospirifer verneuili* and *Homalonotus sp.* also occur, and hard tourmaline rocks akin to those of metamorphic aureoles round the Cornubian granites are common.

Inland the Pebble Beds form an escarpment whose easterly dip slope is marked by barren heathland in contrast with the rich grasslands of the Permian marls; this contrast is commonly sufficient to delineate faults, as for example between Sidmouth and Kentisbeare. As the beds pass northwards there is a decrease in pebble size and in the amount of foreign detritus. This evidence for a southerly origin is supported by H. H. Thomas's work on the heavy detrital minerals occurring in the deposit. However, north of Burles-combe garnet and cassiterite appear in the heavy mineral suite and are ascribed to a westerly current confluent to a drainage system flowing from the north which carried fragments of Carboniferous Limestone, probably derived from the Bristol Channel area and South Wales. The west Somerset deposits are up to 500 ft thick and comprise a variety of breccia types in addition to the pebble conglomerates; the fragments generally include Devonian grits, slates and limestones, quartz and much Carboniferous Limestone.

Upper Sandstones

Overlying the Pebble Beds in the south Devon cliffs is a group of coarse current-bedded red sandstones containing a few quartz and quartzite pebbles in the lowest 100 ft; in turn they are succeeded by more massive sandstones, well exposed between the mouth of the River Otter and Sidmouth. Bones of *Hyperodapedon sp.* have been described by Whitaker (1869) from these beds at Otterton Point.

It is difficult to define precisely the base of the Keuper Series in this region; at Sidmouth the Bunter–Keuper junction has been taken at a hard breccia horizon overlain by pale red and grey sandstones with seams of marl. These sandstones closely resemble recognized Lower Keuper Sandstones in Gloucestershire. At Puriton, near Bridgwater, the formation is 80 to 90 ft thick. In the west Somerset outcrops the proximity of the ancient coastline resulted in the deposition of a series of breccias, conglomerates, marls and a variety of sandstone types which A. N. Thomas (1940) termed ' "Keuper" Sandstones', as opposed to the underlying ' "Bunter" Conglomerates'.

Keuper Marl

The red Keuper marls, near the top of the Triassic, are rarely true marls, but in the main consist of silty mudstones. Thin sandstones occur in the bottom 150 ft of these beds in the cliffs near Sidmouth, and bands of gypsum and rock salt are found in the massive and thickly bedded marls above. The marls constitute the lower parts of the cliffs in the east but near Beer Head they are carried below the beach by a syncline. A boring at Puriton proved 1350 ft of these beds, whence some salt was extracted (pp. 103–4). The coast of west Somerset also shows admirable sections of Keuper Marl, including the passage from Red Marls to Tea Green and Grey marls and upwards into the Rhaetic. West of Minehead the Red Marls change in facies to sandstones, sandy limestones and breccias related to marginal deposition; however, in the Vale of Porlock red and grey marls were deposited away from the coarser scree sediments flanking the basin, indicating a connection between the Porlock and Watchet areas during Red Marls times, a channel which persisted into Liassic times.

In both north and south coast sections grey and green marl bands appear towards the top of the Red Marls sequence and there is a passage upwards into the Tea Green and Grey marls. Gypsum bands are commonly associated with the green bands in the Watchet sections.

Tea Green Marl and Grey Marls

The Tea Green and Grey marls consist of calcareous green shales and mudstones which pass up into grey marls. At Blue Anchor the combined thickness of these beds is 110 ft; on the south coast near Culverhole less than 50 ft are present. The upper part of the Grey Marls (Sully Beds) at Lilstock has yielded the Rhaetic bivalve *Rhaetavicula contorta*, indicating little or no stratigraphic break between the Tea Green Marl and the Rhaetic.

Rhaetic

The transition from Keuper Red Marls upwards into the Grey Marls in west Somerset is best exposed in the coastal sections between Watchet and Lilstock. On the south coast, east of Axmouth, the Red Marls are overlain by a group of unfossiliferous grey and greenish grey marls which have been taken to represent passage beds between Keuper and Rhaetic. At Culverhole, where over 20 ft of Rhaetic strata are exposed, the base of the series was thought by Woodward to lie at a conspicuous horizon of black marl, bounded above and below by whitish marlstones, within the Grey Marls. Richardson (1906; 1911) divided the series into Lower Rhaetic (Sully Beds and Westbury Beds) and Upper Rhaetic (Cotham Beds, Langport Beds—including the White Lias—and Watchet Beds, all of which he considered to belong in the Triassic). He took the base of the Rhaetic at a thin gritty black shale and bone bed separated from the Grey Marls below by a slight non-sequence. At Blue Anchor, near Watchet, the Westbury Beds are underlain by greyish green and yellow-weathering marls referred to the Sully Beds, and the non-sequence and bone bed here lie above the horizon at which Rhaetic fossils first appear in the region. The succession at Blue Anchor is:

		Ft	in
Watchet and Langport Beds	White Lias, laminated shales and limestones	9	10
Cotham Beds	Greenish grey shales and thin shelly limestones with thin Cotham Marble equivalent 4 in from top..	3	10
	Pale limestone with thin brown marl at base	1	6
Westbury Beds	Black shales (limestones in upper 6 ft), fossils including *Rhaetavicula contorta*, *Chlamys valoniensis*, *Tutcheria cloacina* and *Eotrapezium concentricum* ..	23	8
	Ceratodus Bone Bed; massive grit with fish remains resting on thin ripple-marked sandstone ..	–	6
	Black fossiliferous shales and thin, commonly nodular, limestones; occasional sandstones in lowest 4 ft ..	21	10
	Basal Bone Bed ..	–	2
Sully Beds	Greyish green and yellow-weathering marls	13	1
Tea Green and Grey marls ..		110	0

The Cotham Marble or 'Landscape Stone' characteristic of the Bristol area is represented in the Somerset sections by a nodular limestone of Cotham type and in south-east Devon by a pale, slightly pyritic, greenish grey limestone of 'false-Cotham' type. For mapping purposes the base of the Rhaetic has been taken at the bottom of the black shales (Westbury Beds) coincident with the first appearance of marine Rhaetic fossils over much of the region, and the upper limit at the junction of Cotham Marble and White Lias. Interesting sedimentological structures occur in the White Lias in the cliff sections at Charton and Pinhay bays between Culverhole and Lyme Regis.

The limestone and shale beds overlying the Cotham Marble have been associated on lithological grounds with the Blue Lias as part of the Liassic formation, although they are faunally rather distinct. The junction of the Triassic and Jurassic has been taken below the White Lias by some authors and above by others.

Inland exposures in Devon are generally small and few; among the best is Tolcis Quarry, near Axminster. In Somerset there are sections in railway cuttings at Cossington, Charlton Mackrell, Dunball and Langport.

7. Jurassic Rocks

Lower Lias rocks which crop out on the coast of south Devon between Axmouth and Lyme Regis (Fig. 17) afford probably the best large section of these strata in the country. They dip gently eastwards. Some minor folding and a few small faults, generally downthrowing east, are visible in the cliffs and the Liassic beds are overlain unconformably by the Gault and Upper Greensand. The cliffs are so much affected by landslipping as to be locally dangerous. Inland, the valley of the River Axe immediately upstream of Axminster is cut in Lower Lias clays and limestones. There are a few long-abandoned brickpits in the clays, and the White Lias is exposed in a quarry at Tolcis. In 1822 Buckland noted quarries on each side of the Axe valley near Axminster where limestone bands were being worked for building stone and paving stone. Similar workings were formerly active around Staple Fitzpaine, five miles south-east of Taunton, but the limestone is not a durable building stone. Near the Somerset coast Woodward (1893) recorded sections of Lower Lias clays and limestones in a railway cutting and a quarry near Dunball, and over 200 ft of such rocks overlie Rhaetic Beds in cliffs at St. Audrie's Bay, West Quantoxhead. Cliffs of Lower Lias extend most of the way from Watchet to Stert Flats, and boreholes in the Highbridge area have proved up to 400 ft of Lower Lias clays.

Marls, silts and sands of the Middle and Upper Lias crop out in the valley of the River Axe and between Chard and Ilminster.

Lower Lias

The Axmouth–Lyme Regis coastal section shows the following Lower Lias beds:

		Ammonite Zones
Black Ven Marls or Black Marl: shales and cementstones	150 ft	*Echioceras raricostatum* *Oxynoticeras oxynotum* *Asteroceras obtusum*
Lower Black Ven Beds or Shales-with-Beef: shales with fibrous calcite	70 ft	*Caenisites turneri* *Arnioceras semicostatum*
Blue Lias or Lyme Regis Beds: limestones and shales	105 ft	*Arietites bucklandi* *Schlotheimia angulata* *Alsatites liasicus* *Psiloceras planorbis*
White Lias: thinly bedded limestones with marl partings	25 ft	

The White Lias is well exposed in Charton Bay and also in Pinhay Bay, where it has been studied by Hallam (1960). The unweathered limestone is

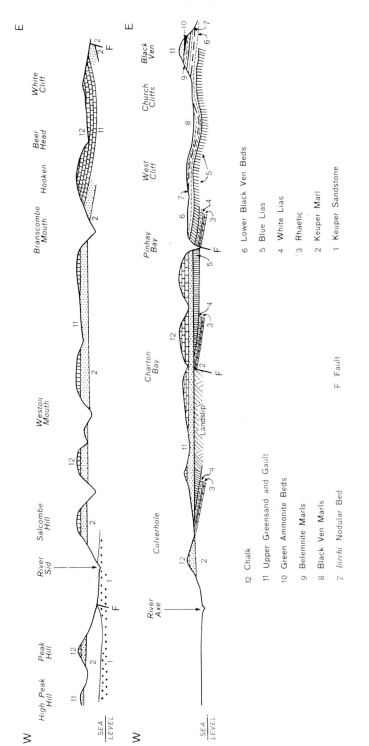

12 Chalk

11 Upper Greensand and Gault

10 Green Ammonite Beds

9 Belemnite Marls

8 Black Ven Marls

7 *birchi* Nodular Bed

6 Lower Black Ven Beds

5 Blue Lias

4 White Lias

3 Rhaetic

2 Keuper Marl

1 Keuper Sandstone

F Fault

Horizontal Scale

Fig. 17. *Cliff section between High Peak Hill and Black Ven*
Mainly from 'Geology of the country near Sidmouth and Lyme Regis', *Mem. Geol. Surv.* 1906

almost pure calcite, but with unusually high strontium content which suggests original deposition as aragonite mud. At the top the Sun Bed, a 2-inch limestone, shows polygonal sun cracks and many U-shaped burrows made by sediment-eating worms or crustaceans. The nature of this bed, local conglomeratic facies within the limestones, and irregular and reddened surfaces, combine to suggest that slight changes in sea level during deposition caused limestone islands to rise above the water in White Lias times. Occasional slight movement of the sea bed caused slumping and a 5-ft limestone near the top of the formation shows intense contortion and brecciation.

The Blue Lias comprises alternating bands of limestone and more or less limy clay or shale, locally pyritic, together with a few layers of lignite. The beds are disposed in anticlines at $\frac{3}{4}$ mile east of Culverhole and at Charton Bay. They are thrown down 40 ft to the west by a fault in a water course at Pinhay Bay and are well exposed in cliffs between Pinhay and Lyme Regis. Limestones are mainly confined to the basal 85 ft of these beds; they are overlain by 16 ft of clay and shale on which rests a prominent $3\frac{1}{2}$-ft band of hard grey marl, Table Ledge, the top bed of the Blue Lias and a useful marker horizon. Both shales and limestones are commonly fossiliferous, yielding mainly ammonites and bivalves. A bed with fish remains occurs near the top of the formation and reptile bones, of *Ichthyosaurus*, *Plesiosaurus* and pterodactyls, have been found.

Immediately east of Lyme Regis an anticline brings up Blue Lias in Church Cliffs. A short distance west of the town the Blue Lias dips east below beach level, permitting easy examination of the overlying Lower Black Ven Beds which consist of shales and marls with bands of impure limestone. Numerous cross-fibre calcite veins ('beef') up to 4 inches thick show cone-in-cone structure and their appearance has given rise to the term Shales-with-Beef. The lowest 40 ft contain most of the impure limestones and mudstones and the highest 30 ft most of the 'paper shales' and 'beef'. Near the top is the *birchi* Nodular Bed, comprising limestone nodules up to 3 ft in diameter which commonly contain good specimens of the ammonite *Microderoceras birchi*, known locally as 'tortoise ammonites' or, where the shell chambers are filled with white calcite, as 'white ammonites'. A few lower horizons yield well-preserved ammonites and bivalves, but in general the fossils are fragmentary, although common.

A foot of shale separates the *birchi* Nodular Bed from the *birchi* Tabular Bed, a 4-inch limestone bed at the base of the Black Ven Marls. These beds were called Black Marl by Lang (1914), a name used by Woodward (1893) to include also the Lower Black Ven Beds. They consist of shales, clays and marls with nodules and a few thin impersistent bands of limestone and cementstone. As in the other Liassic divisions, several of the limestones have been named, such as the (successively higher) Lower Cement Bed, Upper Cement Bed and Coinstone, which forms the Coin Stone Ledge. Ammonites and bivalves are common, but best preserved in the limestones. The Coinstone lies 30 to 40 ft below the top of the Black Ven Marls; indentations on its top may represent borings on an erosion surface and the bed may mark a slight break in sedimentation. The shales above the Coinstone contain a lenticular earthy limestone, the Watchstone Bed or Watch Ammonite Stone, commonly about 1 ft thick and crowded with ammonites.

Lyme Regis is built on a syncline of Black Ven Marls which occasionally slip and cause damage to buildings along the sea front.

The cliffs at Black Ven show pale grey and brown Belemnite Marls overlying the Black Ven Marls, and at the east end of the hill a few feet of the Green Ammonite Beds, the top division of the Lower Lias, are seen beneath the Gault.

The coastal outlier of Lower Lias strata east of Watchet contains well over 500 ft of beds extending up to the *semicostatum* Zone. About 40 ft of limestones and shales with *Psiloceras planorbis* are succeeded by a thick sequence of shales with a few thin limestones, well exposed at the coast north of Kilve. Above are some 60 ft of shales in which limestones become more numerous and more massive upwards (*bucklandi* Zone) and the overlying bituminous shales, about 40 ft thick, with thin limestones yielding *Arnioceras semicostatum*. The beds show many undulations and faults; downthrows are to north and south. A few old kilns attest early attempts to burn limestone for lime.

Middle and Upper Lias

Between Chard and Ilminster the Lower Lias clays pass upwards into bluish grey micaceous clay and silt. The thickness of these basal Middle Lias beds is difficult to estimate but may locally be up to 100 ft. They are overlain by the Pennard Sands, 20 to 50 ft of micaceous silts and fine yellow sands with scattered ferruginous and sandstone doggers and some sandstone beds. The top of the Middle Lias is marked by the Marlstone Rock-Bed, a shelly ferruginous limestone 8 to 12 ft thick. Near Ilminster and Chard this bed is welded to a 1-ft Upper Lias limestone and the combined 'Junction Bed' forms a prominent platform. Shales and limestones of the basal Upper Lias at Ilminster have yielded fish, insect and reptile remains; the limestones contain well-preserved ammonites, including species of *Harpoceras*, *Harpoceratoides*, *Hildoceras* and *Dactylioceras*, which are common in the surface brash of the fields of this area.

8. Cretaceous Rocks

Cretaceous rocks crop out on the eastern margin of the region, where they consist of three broad divisions (Fig. 18): Gault, Upper Greensand and Chalk (with Cenomanian Limestone locally at its base).

In the following pages reference is made to the Continental system of stages and substages: thus the Gault and Upper Greensand belong to the Albian Stage (although locally the Upper Greensand may include strata of Cenomanian age), while the Cenomanian Limestone is equivalent to part of the Lower Chalk of other areas. These strata form the upper part of a highly dissected upland extending from Lyme Regis to Chard on the east, and from Sidmouth to the Blackdown Hills on the west. Most of the high ground is formed by Upper Greensand (Plate 9B) capped by irregular patches of flint and chert gravel, but outliers of Chalk and Cenomanian deposits are preserved locally, particularly on the downthrow side of major faults. The Haldon Hills, south-west of Exeter, consist in part of Upper Greensand and represent the westernmost outlier of Cretaceous strata in the region.

The Gault and Upper Greensand comprise clays and sands of a single formation laid down in water which deepened eastwards. Where the two facies are present together the Upper Greensand invariably overlies the Gault. In the present area the Upper Greensand facies predominates with only a thin representative of the Gault at the base of the succession in the east.

The Chalk crops out over a relatively small area. Inland it caps high ground to the east and west of Chard and forms outliers near Membury and Chardstock, as well as at Widworthy and Offwell east of Honiton; on the coast good sections may be seen in the cliffs between Lyme Regis and Sidmouth (Plate 9A).

On the eastern margin of the area the Cretaceous rocks lie unconformably upon the easterly dipping Lias, farther west they overlie the Triassic, and in the Haldon Hills they rest directly on the Permian. This constitutes the 'Cretaceous overstep', one of the most important unconformities in the stratigraphical column.

Gault

On Black Ven, near Lyme Regis, the Gault consists of dark bluish grey loamy silts and clays with pyritic nodules, a conglomerate with Lias pebbles being usually developed at the base. Palaeontological evidence suggests that it is the equivalent of part of the Lower Gault (*Anahoplites intermedius* Subzone), and the lowermost beds of the Upper Gault (*Hysteroceras orbignyi* and *H. varicosum* subzones). An *intermedius* Subzone fauna has also been recorded from Charton Goyle, east of Seaton.

The Gault is nearly 40 ft thick near Lyme Regis but thins westwards and disappears altogether before the line of the River Axe. To the west of Seaton it is probable that the Gault is represented by the lowest 20 ft of dark green clayey sands of the Upper Greensand succession. Inland the Gault has been

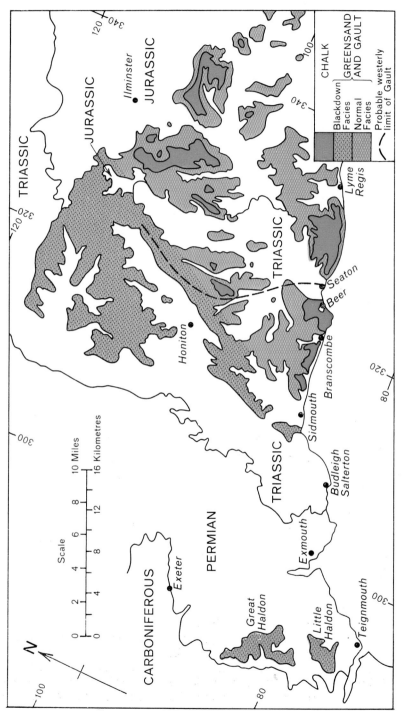

FIG. 18. *Cretaceous rocks of Devon, west Somerset and part of Dorset*

recorded from the railway cutting at the eastern end of the Honiton Tunnel, north-west of Wilmington, but has not been recognized in the Blackdown Hills, where beds of Upper Greensand facies appear to rest directly on the Triassic.

Upper Greensand

The Upper Greensand in Devon is about 150 to 200 ft thick and has been broadly divided into a lower group of glauconitic sands known as the Foxmould and an upper group of calcarenites with bands of chert known as the Chert Beds.

Tresise (1960) recognized two distinct facies, separated by a line extending from the coast at Sidmouth north-east to Yarcombe. East of this line is the 'normal facies', the Foxmould–Chert Beds succession. The 'Blackdown facies' to the west is entirely non-calcareous, comprises fine unconsolidated sands overlain by poorly graded sands with cherts and siliceous sandstones, and is characterized by limonitic staining and well-preserved silicified fossils. The lack of carbonate has been attributed to leaching of an originally calcareous 'normal facies' succession, following the stripping by erosion of the Chalk cover. However, ammonite evidence shows that the two facies are not exactly contemporaneous, and their differences could be original.

Palaeontological evidence of relative ages within the Upper Greensand is meagre. The Foxmould has yielded ammonites (*Mortoniceras spp.*) indicating the *Callihoplites auritus* Subzone of the Upper Albian. The overlying Chert Beds of the coast sections have recently been quoted (Tresise 1961) as belonging to the *Arrhaphoceras substuderi* Subzone, but the ammonite evidence for this is inconclusive. *Hysteroceras varicosum* and other ammonites occur in the lower part of the Blackdown Greensand, and indicate a basal Upper Albian age for at least part of that deposit; the beds with *Hysteroceras varicosum* thus correlate with the upper part of the Gault at Lyme Regis. Early workers considered that the base of the Haldon Greensand (see below) could be equated with a horizon high in the Blackdown Greensand, and that the upper part of the succession entirely post-dated the Blackdown deposits. The little available ammonite evidence tends to confirm this: *Hysteroceras spp.* have been recorded from the lower part of the succession, pointing to correlation with the Blackdown ammonite fauna, while a single ammonite from an unknown horizon determined as the nucleus of a *Stoliczkaia* indicates the presence of uppermost Albian deposits (*Stoliczkaia dispar* Zone) and extends the range of the Haldon Greensand to near the Albian–Cenomanian boundary. It is probable that the highest beds of the Haldon succession are of Cenomanian age.

The 'normal facies'

Perhaps the best section in these beds is that at Whitecliff, near Seaton, where Cretaceous strata are faulted against Keuper Marl. The Foxmould at this point comprises some 85 ft of greenish grey glauconitic sands with courses of calcareous sandstone. The basal beds of dark green sands are overlain by about 15 ft of green sands with very large calcareous concretions which are known as cowstones. A thin nodular glauconitic shelly limestone marks the top of the Foxmould. The overlying Chert Beds are about 65 ft thick and show a wide variety of lithologies including quartzose sands, glauconitic sands, calcarenites, pebble beds, shelly sands and chert in layers and isolated nodules. The uppermost 10 ft or so, W. E. Smith's (1961) 'Top Sandstones', rarely contain chert and are more quartzose than the beds below. Their blocky and indurated upper part passes locally into a 'boulder bed' of subangular joint-controlled blocks, the gaps between the blocks being filled with Lower Cenomanian deposits. Towards the west the Top Sandstones are replaced by calcareous sands largely composed of broken shells, as at Kempstone Rocks near Sidmouth. Inland, in the area around Chard and Crewkerne, the boulder bed is capped by the Calcareous Grit, a thin bed consisting of large rounded quartz grains and sporadic glauconite grains set in a matrix of finer quartz sand, the whole cemented by calcite. This bed is directly overlain by the Chalk Basement Bed and has therefore been regarded in the past as of pre-Cenomanian (i.e. Albian) age; however, palaeontological evidence shows it to be at least in part Lower Cenomanian.

Chert from the Chert Beds has been used by man since prehistoric times, and formed the material out of which the palaeolithic implements found in the Axe gravels were made. West of Chard the beds were formerly extensively worked for road metal at the famous Snowdon Hill quarries, where the main chert-bearing horizon is 30 ft thick. The best section in the Chert Beds, in the cliffs near Seaton, shows sands in which 6- to 9-in-thick chert bands lie a foot or so apart. The main chert-bearing horizon is about 70 ft thick near Seaton, but commonly much thinner elsewhere. The Chert Beds give rise to a good feature with abrupt scarps, as for example Honiton Hill and Buckton Hill. Wherever the Chalk cover has been stripped the Chert Beds have been rapidly disintegrated by weathering; the sand has been washed out and the chert layers have settled down in broken masses, thus giving rise to a considerable thickness of tightly packed angular chert. Further weathering and subsequent redisposition of such deposits has formed the Angular Chert Drift.

Two distinct types of chert are known from the Chert Beds. The 'cored cherts' occur as layers of irregularly shaped porous, cream-coloured nodules, each with a massive black core. The second type of chert occurs as irregular yellowish brown slabby masses which commonly contain geodes lined with mammillated chalcedony or euhedral quartz. In the Seaton area the cored cherts are restricted to the lower part of the Chert Beds, but this does not hold elsewhere. The origin of the cherts remains controversial. Perhaps precipitated silica gel underwent dehydration to porous stone and subsequent conversion into chert by the redistribution of the silica from sponge spicules. Another possibility is the early separation of silica from an original silica gel–calcareous ooze complex. Where chert bands alternate with sands containing nodules of limestone, as in the Chert Beds succession of the coast, it would appear that some kind of two-way movement of silica and calcium carbonate has taken place, although this might have occurred at a comparatively late stage.

The 'Blackdown facies'

The Blackdown facies occupies an approximately triangular area of outcrop defined by Blackdown Hill in the west, Yarcombe in the east, and Peak

Hill near Sidmouth in the south; the Haldon Hills represent a westerly outlier of the Blackdown facies only four miles from Dartmoor.

At Peak Hill the dark argillaceous Foxmould sands are absent, and the succession comprises 30 ft of light grey sands overlain by over 30 ft of grey and buff sands with siliceous concretions. A layer of friable sandy nodular concretions with silicified fossils lies within but near the base of the upper sands.

In the Blackdown Hills the Upper Greensand succession is approximately 100 ft thick and is known as the Blackdown Greensand. About 30 ft of whitish brown indurated sand (apparently the equivalent of the lower sands of Peak Hill) are overlain by highly fossiliferous sands with layers of siliceous concretions. The Blackdown Greensand has long been famous for its fauna of superbly preserved silicified fossils in which bivalves and gastropods predominate but which also includes ammonites and echinoids. The ammonites and nearly all the gastropods are confined to the lower part of the fossiliferous sands, the gastropods commonly occurring in clumps which represent life-associations; the fauna of the upper part of these sands is dominated by bivalves, occurring as isolated and water-worn valves. The beautiful fossils of the Blackdown Greensand were avidly sought by collectors, notable among whom was the Rev. W. Downes whose vast collection is now in the Exeter Museum. Some of the commonest fossils are: *Dimorphosoma calcarata*, '*Murex*' *calcar*, *Tornatellaea affinis*, *Turritella* (*Torquesia*) *granulata*, *Calva* (*Chimela*) *caperata*, *Cucullaea glabra*, *Exogyra obliquata*, *Epicyprina angulata*, *Glycymeris umbonata*, *Nanonavis carinata*, *Paraesa faba*, *Protocardia hillana*, *Pterotrigonia aliformis*, *Venilicardia cuneata*.

In the early years of last century the Blackdown Greensand was extensively exploited for siliceous concretions for the manufacture of whetstones, the best stone lying about 80 ft beneath the surface and being reached by driving long horizontal galleries into the hillsides. The stone as dug was soft and could be shaped, but later it hardened on exposure to air.

Upper Greensand of Blackdown facies forms the highest parts of the Haldon Hills, Great Haldon and Little Haldon, which rise to over 800 ft O.D. and are capped by flint gravels. This so-called Haldon Greensand is up to 270 ft thick on Great Haldon and almost 100 ft on Little Haldon. Somewhat obscured sections may be seen in the pit at the top of Telegraph Hill on Great Haldon and the adjacent excavations for road widening, but the classic section in the old quarry at the head of Smallacombe Goyle on Little Haldon is now badly overgrown. A basal conglomerate with pebbles derived from the underlying Permian breccias is overlain by bedded glauconitic sands with fossils silicified as in the Blackdown Greensand (*Protocardia hillana*, '*Trigonia*' *spp.*, *Venilicardia cuneata*, etc.). The succeeding glauconitic and pebbly sands underlie sands with cherts. The pebbly sands suggest the close proximity of a shore line, as does the occurrence of compound corals and fragmented shells. The fossils are on the whole less well preserved than those from Blackdown and tend to be confined to bands. Most of the fossils come from the 'Haldon coral-bed', a pebbly glauconitic shelly sand in the lower part of the succession which is locally cemented by silica into hard irregular masses. The large foraminiferan *Orbitolina concava* occurs in cherts near the top of the succession.

(A 6414)

A. Chalk cliff east of Seaton Hole, Seaton

Plate 9 *(For full explanation see p. x)*

B. Buckton Hill, Upper Greensand capping Triassic rocks

(A 6400)

(A 9809)

A. Pottery clay workings, Newton Abbot

Plate 10   (*For full explanation see p. xi*)

B. Raised Beach and associated deposits, Widemouth Bay

(A 10369)

Chalk

Lower Chalk and Cenomanian Limestone

The Lower Chalk is represented in the region by thin and condensed deposits. W. E. Smith's (1957; 1961) detailed accounts of the Cenomanian strata of south-east Devon distinguish three different facies: the Lower Chalk facies, the Cenomanian Limestone facies and the calcareous sand or Wilmington facies.

The Lower Chalk facies occurs in the Membury outlier and can be examined in the large chalkpit alongside the narrow lane leading to Furley. The Chalk Basement Bed, of phosphatized limestone pebbles and fossils set in a matrix of sandy limestone, rests on eroded Calcareous Grit of the Upper Greensand and is overlain by 3 to 4 ft of glauconitic chalk. Above lie over 50 ft of soft white well-bedded chalk with a few layers of porcellanous siliceous nodules in the lower beds. The lower part of the Chalk Basement Bed yields Lower Cenomanian ammonites, including species of *Hyphoplites*, *Mantelliceras* and *Schloenbachia*, preserved in sandy limestone; ammonites from the higher part of the Basement Bed are phosphatized, and include species of *Calycoceras* indicating a Middle Cenomanian age. Fossils other than *Inoceramus pictus* are rare in the white chalk, but a single specimen of an Upper Cenomanian *Schloenbachia* has been recorded (Smith and Drummond 1962).

The Lower Chalk facies is not confined to the Membury outlier, but extends into Somerset and west Dorset. A somewhat similar succession to that just described is seen in the Snowdon Hill quarries to the west of Chard; at this locality the Basement Bed is richly fossiliferous and has yielded large numbers of well-preserved ammonites as well as other fossils.

In the coastal outliers the Cenomanian is represented by thin indurated sandy limestones known collectively as the Cenomanian Limestone. This is generally from 1 to 4 ft thick, but locally, as at the Hooken and Beer Head, thicknesses of almost 30 ft have been recorded; at the western end of Charton Cliff the limestone is absent and Middle Chalk rests directly on the Upper Greensand. Smith (1961), following the work of Jukes-Browne (1903), recognized three main divisions of the Cenomanian Limestone, discontinuous and present together only in thick developments of the formation as at the Hooken. The basal hard white shelly limestone (Division A, 0 to 18 ft) comprises a lower part with abundant grains of quartz and the coral-like bryozoan *Ceriopora ramulosa*, and an upper shelly limestone with relatively little quartz and an erosion surface at the top. Division B (0 to 5 ft) is firmly welded on to the underlying rock and consists of hard white sandy glauconitic limestone, the sand and glauconite commonly being concentrated in lenses; phosphatized pebbles of limestone are numerous, particularly towards the top. The top of Division B also is an erosion surface and is marked locally by a thin phosphatized pebble bed. Division C (0 to 7 ft) is entirely local in distribution and forms a sandy base to the Middle Chalk; it is white and chalky with much sand and glauconite and is characterized by the abundance of the rhynchonellid *Orbirhynchia wiestii*. Where Division C is absent chalk of the *Inoceramus labiatus* Zone of the Middle Chalk rests directly on the erosion surface at the top of Division B.

The Wilmington facies is a coarse calcareous sand consisting of grains and small pebbles of quartz set in a chalky matrix. Where a Middle Chalk cover has prevented extensive decalcification and weathering a variable thickness of hard calcareous sandstone ('grizzle') is commonly present near the top and is separated from the Middle Chalk by a thin glauconitic sandy limestone resembling Division B of the Cenomanian Limestone. The facies is well exposed in the large sandpit opposite the 'White Hart' in Wilmington, where the sands are about 40 ft thick and rest on a hard glauconitic sandstone with quartz pebbles. The 'grizzle' is highly fossiliferous and large numbers of fossils can be collected from weathered lumps and piles of screened material on the floor of the pit. Echinoids are particularly common and many different genera are represented. Bivalves such as *Chlamys* (*Aequipecten*) *aspera* and the rhynchonellid brachiopod *Cyclothyris difformis*, also characteristic of the fauna, have their shells partly replaced by the chalcedonic mineral beekite. The age of the 'grizzle' is given by comparatively poorly preserved Lower Cenomanian ammonites belonging to the genera *Hyphoplites*, *Mantelliceras*, *Schloenbachia* and *Turrilites*. The Wilmington facies thins north of the village to less than 20 ft at Hutchin's sandpit, where the sands rest on a shell-bed containing abundant *Trigonia* and sporadic, but well preserved, Lower Cenomanian ammonites. Sands of this facies are also developed in the Beer district, in an outlier where most of the Cenomanian deposits are Cenomanian Limestone; a good section is seen in a pit on the south side of Bovey Lane.

The relationship between the three facies of the Cenomanian is complex. The Wilmington sands are at least in part Lower Cenomanian, while derived ammonites (*Protacanthoceras*) occurring at the base of the overlying Middle Chalk provide evidence of the reworking of Upper Cenomanian sediments. Since both Lower and Middle Cenomanian ammonites occur in the Chalk Basement Bed at Membury the overlying chalk with siliceous nodules must post-date the Lower Cenomanian Wilmington sands. Ammonite evidence shows Division A of the Cenomanian Limestone facies to be Lower Cenomanian, Division B to be of uncertain age but not older than the upper part of the Lower Cenomanian, and that a big faunal break occurs between Divisions B and C. The phosphatized pebble bed locally present at the top of Division B records a complex history of sedimentation and reworking. It contains large numbers of ammonites which have undergone varying degrees of phosphatization, and includes some (e.g. *Tarrantoceras rowei*) which indicate the highest beds of the type Cenomanian; other ammonites such as *Euomphaloceras* point to still higher horizons. Smith (1961) suggested that Division C of the Cenomanian Limestone should be renamed the *Orbirhynchia* Band, and that the term Cenomanian Limestone should be restricted to Divisions A and B. The exact stratigraphical age of Division C is uncertain, but the occurrence of the belemnite *Actinocamax plenus*, as well as glauconitized rolled *Metoicoceras*, suggests correlation with the *plenus* Marls of the standard southern England Chalk succession, which are basal Turonian.

Smith (1957; 1961) established that Cretaceous sedimentation was influenced intermittently by axes of uplift. These appear to be related to periclinal structures and are orientated in an approximate north–south direction. Two particularly important axes are located near Branscombe

Mouth and along a line joining Beer Beach and Little Beach. The thickest Cenomanian successions accumulated in depressions between axes. The limestone of the Hooken–Beer Head and the Wilmington sands facies of Bovey Lane are thought to have accumulated in similar but separate depressions.

Middle and Upper Chalk

The remarkable variations in thickness, consequent upon lateral variation and overlap against tectonic axes of the zones of the Middle and Upper Chalk exposed on the coast, are shown in feet in the following table (based on Rowe 1903), although it should be noted that several of the figures are estimates and the Pinhay sections are now considerably overgrown.

	Zone	Berry	Brans-combe	Beer Head–Hooken	Beer and Annis' Knob	White-cliff	Pinhay
Upper Chalk	*Micraster cortestudinarium*				30 exposed		50 exposed
Upper Chalk	*Holaster* (*Sterno-taxis*) *planus*			50	60	45–50 estimated	40
Middle Chalk	*Terebratulina lata*		35 exposed	156 exposed	90 estimated	70 estimated	72 exposed
Middle Chalk	*Inoceramus labiatus*	80		16	26	28	60

In the present area the whole of the Middle Chalk and the two lower zones of the Upper Chalk are present. The Middle Chalk has a very restricted outcrop. Inland it occurs as outliers at Lady's Down between Crewkerne and Chard, at Cricket St. Thomas, Combe St. Nicholas, Membury and Widworthy; on the coast it may be seen in the cliffs at Pinhay, Whitlands, Beer Harbour and Beer Head.

The *Inoceramus labiatus* Zone is normally flintless and nodular and there is no Melbourn Rock at its base. The lowest bed is a hard rough gritty limestone with grains of quartz and glauconite and is overlain by a bed of chalk with hard yellow nodules. Locally the base of the Middle Chalk is marked by Division C (*Orbirhynchia* Band) of the Cenomanian Limestone. Another local development near the base is the so-called Beer Freestone, up to 13 ft thick, which is a granular limestone made up of comminuted *Inoceramus* and other shell debris. A similar freestone known as the Sutton Stone occurs near the base of the zone in the Widworthy outlier. The Beer

Freestone was formerly extensively exploited in quarries to the west of Beer but demand for the stone is now small. When freshly quarried it is soft enough to be cut and dressed with steel saws, but exposure to air produces hardening. The stone is very suitable for carving, especially for interior work. It has been used in the construction of a shelter overlooking Beer Beach and may be seen in the interiors and exteriors of churches at Ottery St. Mary, Honiton and Axminster; the stone was also used in the cathedral at Norwich.

The sandy Division C of the Cenomanian Limestone yields rolled Lower Turonian ammonites (*Metoicoceras*); a somewhat higher horizon is indicated by rolled ammonites (*Neocardioceras*) from a pebble bed found near the base of the Middle Chalk at Haven Cliff. The main mass of the *Inoceramus labiatus* Zone is very fossiliferous, the following species being particularly characteristic: *Orbirhynchia cuvieri, Cardiaster pygmaeus, Dixonia dixoni, Peroniaster nasutulus* and *Inoceramus labiatus*. A noteworthy feature is the presence of *Micraster*, which is exceptionally rare in this zone in other areas. This normal *labiatus* Zone fauna is found to the east of Branscombe. To the west, however, the *labiatus* Zone chalk contains spines and plates of cidarids in profusion, together with ossicles of crinoids and asteroids, and the normal fauna is absent. In the coastal outliers between Branscombe Mouth and Berry Cliff 20 ft of this flintless chalk with an abnormal *labiatus* Zone fauna are overlain by 60 ft of flinty chalk. The latter resembles the chalk of the *lata* Zone, but it contains the same abnormal fauna as does the chalk below and must be placed in the *labiatus* Zone.

The *Terebratulina lata* Zone in Devon consists of massive marly chalk which differs from the equivalent beds of other areas in being very flinty, with nodular flints arranged in well-marked courses. Two flintless marly bands in the sections around Beer provide excellent marker horizons. The greatest thickness recorded is 156 ft at the Hooken, which indicates that tectonic control of sedimentation continued during deposition of the Middle Chalk. The rich fauna includes *Conulus subrotundus, Dixonia dixoni, Holaster* (*Sternotaxis*) *planus, Micraster spp.* and *Peroniaster nasutulus;* the zonal index fossil occurs in greater profusion and grew to a larger size in the *lata* Zone of Devon than elsewhere.

The Upper Chalk crops out on the coast between Pinhay and Rousdon and at Beer, and is also found in two small outliers north of Chard. The most accessible section is provided by the bluff known as Annis' Knob, by the side of the cliff path between Beer and Seaton. Beer Head forms the westernmost outcrop of Upper Chalk in England.

The chalk of the *Holaster* (*Sternotaxis*) *planus* Zone is nodular in contrast to the marly chalk below, and contains numerous courses of nodular flints; at Annis' Knob the nodules are formed of partially silicified chalk. The fauna is rich in echinoids, particularly the zonal index and *Micraster praecursor*. Near the base about 12 ft of yellowish white nodular rocky chalk with scattered phosphatized lumps are possibly equivalent to the Chalk Rock of other areas, although the usual *Hyphantoceras reussianum* fauna of moulds of aragonitic fossils is represented only by sporadic occurrences of '*Turbo*' *gemmatus* and *Bathrotomaria perspectiva*.

The chalk of the overlying *Micraster cortestudinarium* Zone is also nodular but includes one or more marl seams.

The former presence of higher zones of the Upper Chalk is shown by the fossils from the flints of the Haldon Gravels, which cap the Haldon Green-sand. These fossils have come from zones up to and including the *Belemnitella mucronata* Zone. They are usually exceptionally well preserved and extensive collections can be seen in the museums at Exeter and Torquay.

9. Tertiary and Quaternary

Sands and gravels of possible Tertiary age and marine origin occur at 800 ft O.D. on the Haldon Hills and 350 to 500 ft O.D. at Polcrebo, Canon's Town, Crousa Common, St. Agnes and Orleigh Court (Fig. 19). Clays at Filleigh have been dug from alluvial deposits, while sands, clays and gravels at St. Erth have been assigned to the Pliocene; recently, however, a Pleistocene age has been suggested for the St. Erth deposits. Boulder clay occurs at Fremington and on St. Martin's in the Scillies, and the huge gneissic Giant's Rock at Porthleven was presumably deposited by an ice sheet or ice floe.

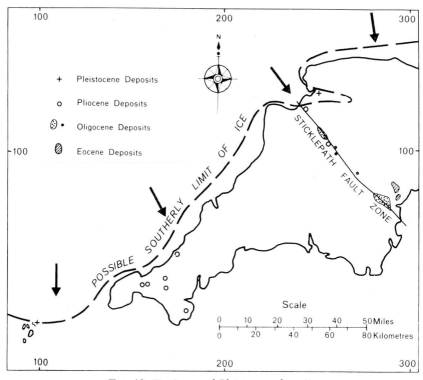

Fig. 19. *Tertiary and Pleistocene deposits*

Eocene

It has long been known that the various Eocene beds become coarser and more gravelly westwards from Hampshire into Dorset. They represent sedimentation in deltas, lagoons and shallow near-shore sea water. Two different formations, the Reading Beds and the Lower Bagshot Sands, pass westwards into true subangular river gravels; but in the Tertiary outliers west of Broadmayne the Reading Beds are overlapped by the Bagshot gravels which there rest on the Chalk. The constituents of these gravels indicate derivation from exposed Mesozoic and Palaeozoic sediments, and also granite, to the west.

The gravels on Great Haldon Hill and Little Haldon Hill are possibly of Bagshot age. Both hills are flat topped and each is capped by a thick sheet of coarse, partially rolled gravel, mixed with seams of granitic sand and whitish clay. Much of the gravel consists of more or less worn masses of Chalk flint, commonly of large size, and although the nearest Chalk outlier now lies 15 miles away to the east the Chalk cover may have been more extensive in Eocene times. The other coarse constituents comprise subangular Greensand chert and Palaeozoic stones, the latter perhaps more usually derived from the Permian breccias than directly from the Palaeozoic rocks. The matrix consists to a great extent of rough granitic sand, mixed with whitish clay derived probably from the decomposed feldspar of the granite.

Locally the gravel reaches 30 to 40 ft in thickness, but it has been thoroughly decalcified and therefore is without fossils except moulds in flint of Cretaceous forms.

Oligocene

The clays of the Bovey Tracey region have long been known and worked (Plate 10A). They are associated with sands, gravels and lignites in a basin which lies on the line of the great Sticklepath–Lustleigh tear fault. Sagging within the fault zone is thought to have allowed the accumulation of a considerable thickness of sediments. A 456-ft borehole at Heathfield and a 667-ft borehole at Teigngrace failed to reach the base of the Oligocene deposits. Bott and others (1958) interpreted a small negative gravity anomaly over these deposits as indicating that the sediments lie in a basin whose depth is small in comparison with its width and whose base is probably at about 200 metres below sea level, not far below the bottom of the Teigngrace borehole. However, more recent geophysical surveys indicate that the basin is about 4000 ft deep (Fasham 1971).

The lake, possibly of brackish water, in which most of these sediments were laid down was at least ten miles long and about four miles wide. It received the drainage of several mountain torrents from Dartmoor which pushed out extensive deltas of granitic sand and mud mixed with masses of wood and peat that later formed lignite. Locally the Bovey Beds become very coarse and gravelly. From top to bottom the lignites contain twigs of *Sequoia couttsiae*, in many cases in enormous profusion, and it is probable that the bulk of the peat is formed of this tree, although *Nyssa* and *Cinnamomum* are also abundant. These trees formed water-logged rafts of vegetation of the nature of 'pine rafts', but no suggestion of a swamp deposit has been found; true swamp vegetation, despite a long search for it, was represented only by a single scale of a cone belonging to the swamp-cypress *Taxodium distichum*, a tree which grows equally well on land.

The closest parallel to these beds is the brown coal deposits of Germany, which are generally accepted as upper Oligocene in age, though some consider them to be Aquitanian (lowest Miocene).

On the eastern side of the basin numerous exposures in the lower part of the Bovey Beds show the following composite section:

1. Sandy beds with clays and some lignite.
2. 'Potting Clays' or 'Whiteware Clays'—used in the manufacture of high-grade white earthenware—with associated lignitic clays and lignite seams.
3. 'Stoneware Clays'—a group of siliceous clays used in the manufacture of coarse ware such as drain-pipes, tiles and bricks; some local development of sandrock; no lignite.
4. 'Pinks'—siliceous clays with grains of iron oxide; no lignite.
5. Basal sands and gravels with some clay; no lignite.

The lignite has been used as a fuel in the potteries; it yields up to 4 per cent montan wax. The main seam, known as the 'Big Coal', is locally more than 10 ft thick.

Fluviatile and lacustrine deposits, known as the Marland Beds and similar to the Bovey Beds, occur at Petrockstow in north Devon in a sedimentary and slight topographic basin about $\frac{1}{2}$ to $\frac{3}{4}$ mile wide and $4\frac{1}{2}$ miles long. The basin lies on the line of a north-westward extension of the Lustleigh–Sticklepath Fault. The top soil is a Quaternary river deposit of stony brown clay 5 to 10 ft thick containing numerous large boulders, and below it lie white plastic unctuous clay, stoneware clays and argillaceous sands, with thin carbonaceous layers. Irregular layers of lignite from a few inches to a foot thick occur. The beds are worked opencast and along drifts driven at different levels from inclined shafts, the principal products being blue ball-clay and stoneware clay.

Bott and others (1958) deduced from a small negative gravity anomaly that the sediments of the Petrockstow basin were at least 245 metres thick and an Institute of Geological Sciences borehole sunk in 1966–7 proved that presumed Oligocene deposits rest on Carboniferous rocks at a depth of 2169 ft.

Small deposits of silt, sand and gravel of possible Oligocene age have recently been mapped near Jacobstowe and Exbourne and near the edge of the granite north of Chagford.

Numerous erosion platforms are thought to have been recognized in south-west England but most are local and of small extent. Of the three main surfaces, the highest, which now lies at about 1000 ft O.D. and is best seen on the granite around Moretonhampstead, may be of Oligocene age.

Miocene

The Alpine earth movements raised Britain above the sea and no Miocene deposits are known. South-west England was sheared by several north-west –south-east wrench faults. Only slight warping of the erosion platforms has been noticed and this has led to the suggestion that even the 1000-ft surface may post-date the orogeny and be of late Miocene age. However the evidence for Alpine folding in the region is slight and a platform cut in granite could well have undergone uplift without evident surface warping. A pause in emergence of land permitted the cutting in Miocene times of a second, lower, marine platform, which was then raised above sea level and is now at about 750 ft O.D. Simpson (1964) considered it unlikely that the sea had ever stood at much over the present 400-ft level in Devon since the Alpine movements.

Pliocene

Spread of the late Miocene and Pliocene sea produced an outline of the British Isles roughly resembling the present shape. An erosional platform was cut by this sea, and its subsequent emergence may have been due partly to uplift and partly to retreat of the sea as water became frozen into the northern ice cap. This platform is now at about 430 ft O.D. and has been traced from end to end of Cornwall and in parts of Devon. It backs against a degraded cliff which locally cuts obliquely across geological boundaries.

Sands and clays at St. Erth, now concealed, are located at about 100 ft O.D. in a valley cut in the platform and estimates of their age have varied from Eocene (Milner 1922) to Pleistocene (Mitchell 1966). About 4 ft of Head overlie some 20 ft of sediments which locally show the following downward sequence:

> Yellow sand
> Coarse ferruginous sand ('growder')
> Yellow sand
> Blue clay with fossils
> Quartz pebbles
> Fine quartzose sand
> Coarse ferruginous sand

The clay beds are variable in thickness and sporadic in occurrence but have yielded many species of marine mollusca and foraminifera in a fauna of Mediterranean affinities. Theories have been advanced in which some or all of the clays are considered to be boulder clays, but although some disturbance of the beds has been noted, this is no more than could be explained by hill creep. Carriage by ice over any considerable distance would probably have largely destroyed the stratified nature of such sediments, or at least caused extreme contortion, and until the beds are again exposed and yield new evidence they are best considered as a Pliocene marine deposit. Many of the molluscs probably lived at a depth of about 60 ft, but littoral shells are absent, as are also those which inhabited sheltered creeks and estuaries. Reid (1890) suggested deposition of the clay at 250 to 300 ft depth, which might place the land 400 ft lower than it is today, a figure approximating to the amount of uplift of the plain of erosion. Recent examination of foraminifera, however, has led Mitchell (1966) to the view that the dominant forms suggest a depth of less than 65 ft while the large number of species suggests a depth of 80 to 330 ft.

Around the mining region near Camborne the 430-ft shelf is conspicuous, the 'island' of Carn Brea rising from it, while St. Agnes Beacon with its base skirted with sand and clay can be seen 'floating' above the general surface of the north coast of Cornwall from numerous view points. The clays of this area were long dug for 'candle clay', used in the mines for fixing candles at convenient points, and pottery clay for making coarse pottery and bricks. At the base of these beds the old sea-beach is locally exposed, and among pebbles of igneous rocks, vein quartz and sediments, nuggets of tinstone have been found. Conspicuous also is the old cliff at Tintagel and Boscastle, while the inland extension of the platform forms the dreary Goss Moor and the land above Luxulyan valley. Elsewhere its sands have been dug, especially at Crousa Downs, near Coverack, and on Polcrebo Downs, near Crowan.

At Orleigh Court, west of Bideford, a deposit of sand and flints with many derived Chalk fossils is about ¾ mile long by ¼ mile wide. It resembles a beach deposit laid down on the 430-ft platform and is probably of Pliocene age. A small patch of possibly Pliocene gravel lies just north of Hatherleigh.

Uplift of this plateau led to rapid overdeepening of the river valleys, and the characteristic steep-sided gorges of Cornwall and Devon resulted from this incision. St. Nectan's Kieve, the Rocky Valley, Lydford Gorge, and Lustleigh Cleave are well-known beauty spots produced by this agency.

Pleistocene and Recent

Marine erosion and subsequent uplift have produced an uncertain number of raised wave-cut platforms of which only those bearing beach deposits may be termed raised beaches. A higher one or group lies at 50 to 65 ft O.D. and is best preserved at Penlee Quarry, near Mousehole; the ancient cliff is there seen with coarse beach deposits banked against it. Strand lines at 50 ft have been noted near Padstow, Marazion and Trebetherick Point.

A number of possibly separate lower beaches between 5 and 25 ft O.D. may best be regarded as one (Plate 10b). They closely parallel the present shore. Where the land slopes gently to the sea this lower shelf widens out into a broad platform with a small bluff at about 15 to 20 ft marking the old cliff. At numerous localities the old beaches are preserved. The beach deposits vary from sand to boulders and are up to 20 ft thick. Locally they show some cementation by dark brown and black oxides of iron and manganese and they are commonly overlain by Head. A composite sequence of Pleistocene and Recent deposits in Cornwall and Devon is as follows:

Recent:	Submerged forest, alluvium, modern beaches
Devensian:	Head
Interglacial:	5- to 25-ft raised beach, old blown sand, 1st river terrace
Wolstonian:	Boulder clay, erratic pebbles on Lundy, older terraces
Interglacial and Anglian:	Arrival of ice-rafted coastal erratics, earliest terraces
Cromerian:	Shore platform cut

The beach deposits, which in north Devon include and overlie erratics from western Scotland, are overlain by false-bedded sand which may be blown sand although it was thought by Dewey (1935) to be sub-aqueous. Possibly the large coastal erratics arrived during the Anglian glaciation and the succeeding interglacial period. The next (Wolstonian) glaciation brought the Fremington boulder clay, which also contains Scottish erratics. This clay occurs between Barnstaple and Instow. It is up to 78 ft thick and includes lake clays which have been dug for pottery making. The equivalent boulder bed, well seen at Trebetherick Point, may be in part an outwash gravel. The raised beach has yielded a temperate fauna and is well exposed in Fistral Bay, Newquay, and at Saunton. Stephens (1961) assigned it to the earlier interglacial but it could scarcely have survived the passage of Wolstonian ice (Edmonds 1972).

Head is usually regarded as rock debris which has moved downhill by solifluxion in periglacial conditions, although such hill creep has, of course, also taken place in Recent times. In the field it is commonly impossible to distinguish clearly between rock waste which has so moved and that which has weathered *in situ*. Head in south-west England is typically about 10 ft thick, but may be up to 100 ft.

The rise and fall of sea level in Pleistocene times as the northern ice cap receded or advanced, and the consequent changing base levels of the rivers, contributed to the formation of river terraces of silt and gravel. Seven such terraces have recently been mapped in the valley of the River Taw. Regrading of the rivers following melting of the ice resulted in the accumulation in the lower reaches of the valleys of alluvial silt, clay and peat, locally up to 120 ft thick. Fossil remains of forests which grew upon and were engulfed by such deposits are commonly exposed by storms and tides; examples occur at the mouth of the Char, in Tor Bay and Mount's Bay, in the Hayle and Camel estuaries, and at Porlock, Minehead and Stert Flats.

Deposits of Recent times include river alluvium, blown sand and modern beaches. Spits have formed at Dawlish Warren, where an inner and an outer (diminishing) spit are separated by Greenland Lake; at Teignmouth, where a permanent spit (Denn Point) extends south-west into the estuary and a temporary spit extending north-east from The Ness undergoes cycles of building and destruction; and at Northam, where the pebble spit built out northwards from Westward Ho! is being eroded and rolled inland. Barriers of shingle have formed at several places along the south coast, as near Slapton and Helston.

Blown sand is widespread around the coast, for example at Dawlish Warren, Par Sands, Whitesand Bay (Land's End), St. Ives Bay, Widemouth Bay and in the Scilly Isles, particularly on Tresco and St. Martin's. At Braunton Burrows sand-dunes are being reclaimed by the planting of marram grass. St. Piran's Church in Cornwall was engulfed by blown sand about 800 years ago. It reappeared in 1800 and has since been encased in concrete. Peat has developed on waterlogged moorland where temperatures have remained too low for bacterial action to break down dead plants. Inclusions of bog trees, preserved birch, pine and oak, are not uncommon. On high Dartmoor the peat is locally up to 20 ft thick; it is now actively growing only at certain springs or where drainage has become choked, and taken as a whole the peat cover is wasting away. Blanket peat is less common on the Cornish granites, which are lower and drier, but occurs at about 1000 ft O.D. on Bodmin Moor.

The drainage of south-west England was probably superimposed from a Cretaceous cover and generally ran south and south-east except where radial streams had developed on high ground such as Dartmoor which had remained uncovered by the Chalk Sea. The courses of subsequent streams from east and west were later governed by the trend and nature of the Palaeozoic rocks when these became exposed, and diversions were probably caused by ice in glacial times. Where later tributaries cut through resistant rocks across the strike they commonly followed lines of weakness such as faults or major joints. Further complications have been introduced by river captures. The rivers Exe and Tamar drain southwards across almost the whole width of the peninsula but it is predictable that one of the streams flowing to Bude Haven will eventually capture the headwaters of the Tamar. This will produce a loop rather like the present River Camel, which may once have drained into the River Fowey. Similarly, the upper Torridge may have continued eastwards to join the Yeo and been fed by now-reversed streams in the positions of the present lower reaches of the Torridge and the Taw.

Caves and Early Man

Several caves in south Devon have become classic on account of the remains of extinct animals in association with Palaeolithic implements of several types and ages that have been found in them. The principal cave is Kent's Cavern, which lies in the Wellswood district on the west side of a valley leading to Meadfoot Sands (near Torquay), and it was evidence from there which raised some of the first serious objections to the theological dogmas of the Creation and the age of the world. A history of its exploration has been given by Kennard (1945).

The materials of the cave are indicated in Fig. 20. The Black Mould forms the top layer; in it have been found antiquities ranging in age from Neolithic to medieval, associated with bones of domestic animals and of seal, fox, badger, brown bear and long-horned ox.

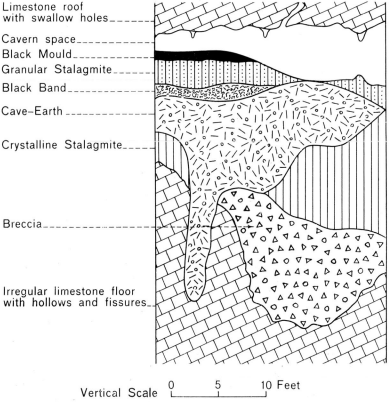

Limestone roof with swallow holes
Cavern space
Black Mould
Granular Stalagmite
Black Band
Cave–Earth
Crystalline Stalagmite
Breccia
Irregular limestone floor with hollows and fissures

Vertical Scale   0   5   10 Feet

FIG. 20. *Diagrammatic section of deposits in Kent's Cavern*

The Granular Stalagmite is up to 3 ft thick; it originally extended nearly all over the cavern but was absent from the extreme western parts and from the large chamber opposite the southern entrance. Embedded in it are bones of brown bear, mammoth, hyaena, woolly rhinoceros, horse, fox and man, and a few flint flakes and cores too indefinite for classification.

The Black Band comprises the highest part of the Cave-Earth, and was so called on account of its dark colour which was due to the presence of burnt bones and charcoal. It is of only local occurrence and probably represents a series of hearths, but it is of archaeological importance because of the artifacts found in it. These included a bone needle and an awl, uniserial and biserial horn harpoons, end-, hollow- and keeled-scrapers, and gravers and points with battered backs and some with a tendency to geometric forms.

The Cave-Earth, though unstratified, shows by the rolled condition of the bones which it contains that it was once washed by water. It is mainly a reddish brown loam containing angular lumps of limestone and some of grit; sporadic blocks of the underlying Breccia and Crystalline Stalagmite occur. It covers much of the cave floor and occupies fissures and hollows, in one spot to a depth of 23 ft, but in parts of the south-western galleries it dies out.

A great number of bones and teeth of extinct animals gnawed by hyaena are mixed with the Cave-Earth and it is probable that the cave was once a hyaena-den. In addition to abundant remains of hyaena, those of horse, woolly rhinoceros, cave lion, cave bear, grizzly bear, reindeer, stag and bison were also common; rather rare were bones of mammoth, wolf, brown bear and beaver. Many other species were found, including glutton, cave pika and sabre-toothed tiger.

The Crystalline Stalagmite is harder and denser than the Granular Stalagmite and locally it attains a thickness of 12 ft, but it dies out altogether in parts of the cave.

The Breccia consists of rounded, subangular and angular pieces of red grit in a sandy matrix, with some fragments of limestone and up to 50 per cent of bones and teeth, principally of bear, cemented into a hard rock. Rolled implements of Chelles type have been found at all levels in the breccia, but they were not numerous.

The fauna of the Cave-Earth is late Pleistocene and the implements recovered represent four periods at least. This mixture of types is probably due to the samples having been collected in foot-levels, which do not correspond with stratification. Pengelly, however, left in his journals the exact location of all the finds and from his notes and the implements Miss Dorothy A. E. Garrod (1926) was able to attempt a correlation of the artifacts. The oldest series is the Mousterian, which includes a number of heart-shaped hand axes and some well-made side-scrapers and large rough flakes all of chert. The next group consists of scrapers of Middle Aurignacian types similar to those from Paviland, a bone pin, and some flint and chert flakes. The third group includes points comparable with proto-Solutrian types, but not many flakes or cores were found. The fourth group contains the largest number of implements and flakes and belongs to the end of Palaeolithic times.

The majority of the implements come from the Black Band. They can be dated by their association with a reindeer antler harpoon that has a single row of barbs and one with two rows of barbs, and some needles of bone. The industry as a whole has none of the characters of the classic Magdalenian, for gravers are very rare and the leading form is the *dos rabattu* point, but the bone and antler implements are typically Magdalenian.

Of the other caves in the limestone, Tornewton Cave, near Torquay, contains a long sequence of Pleistocene strata.

No Palaeolithic implements have been recorded from the river deposits of Cornwall and west Devon although in east Devon they occur in great profusion in the gravels of the River Axe, at Broom, near Axminster. These gravels consist chiefly of Greensand chert, flint and quartz, and blocks of schorl-rock are also met with. They have long been worked either for ballasting the track of the railway or for the roads. They occur in the form of flattened deltas up to a known thickness of 45 ft; the gravel below that depth is flooded and appears to extend below the level of the river. There appear to be two implementiferous levels, one yielding pointed hand axes and the other ovate and cordate forms and large flakes. Many of the implements are large and beautifully trimmed and some of the ovate hand axes show the reversed S twist on their edges. They may be assigned to a late Acheulian period in form and technique. In the gravels of Kilmington, which extend to a depth of 80 ft, rudely chipped cherts have been found.

10. Economic Geology

The two most important industries in south-west England are agriculture and tourism which, in Devon, Cornwall and the Scilly Isles, employ about 63 000 and 26 000 people respectively. Quarrying and mining in Devon and Cornwall are summarized by the following statistics, mainly for 1972:

| | Production (1000s tonnes) | |
	Devon	Cornwall
Tin (metal)		3·3
Tungsten (metal)		0·002
China clay		2961
China stone		53
Potter's clay	(1967) 520	
Other clay and shale	(1968) 208	
Slate		24 (1971)
Limestone.	4554	
Sandstone	1202	
Igneous rock	(1970) 759	1923
Sand and gravel (including marine) . .	1875	824

Mineralization and Mining

Compared with the numerous derelict engine houses and scattered mine dumps so characteristic of the Cornish scene, only four currently productive tin mines reflect the intense mining activity in the region during the 18th 19th centuries (Plate 11A). Over the years ores of tin, copper, lead, zinc, silver (mainly as a by-product of lead), arsenic, antimony, iron and manganese, together with lesser amounts of tungsten, cobalt, nickel, uranium, baryte and fluorspar, have been won from the mineralized areas associated with the granitic intrusions of south-west England. Bismuth, molybdenite and gold also occur, though in very small quantities.

It is believed that cassiterite (tin oxide) has been exploited in Cornwall and Devon since the Bronze Age (1800 B.C.), when the ore was probably obtained solely from alluvial and eluvial gravel deposits. During the Roman occupation a more extensive development of the ore deposits took place. The Romans, already with experience of mineral veins and mining techniques, possibly introduced underground mining to the region and it has been suggested that Ding Dong Mine in the Land's End peninsula and surface excavations near Breage and on Exmoor date from this period. By medieval times silver had also become an economic product in the region; argentiferous galena is thought to have been worked in the 13th century at Combe Martin and Calstock. The output of tin rose appreciably in the 17th century, possibly allied to pewter manufacture. The Industrial Revolution brought a demand for iron ores and many of the South Wales foundries were supplied by the hematite mines of the Exmoor area; copper, manganese, zinc and arsenic were also systematically mined from this period onward.

In the past the major mining districts, mainly for tin and copper ores, were: around Camborne and Redruth; St. Just-in-Penwith; a broad belt between St. Ives and Helston (Fig. 21); St. Austell; and the areas north of Callington and around Tavistock. In 1912 there were 70 active mines in Devon and Cornwall, 45 of them tin producers; several mines continued operating until the late 1930s, for example Basset and Grylls (Porkellis) Mine and Wheal Reeth, while others, particularly in the St. Ives, St. Day and east Cornwall areas, were worked for brief periods during the 1939–45 war. Recent years have seen the closure of the Bridford Barytes Mine (in 1958), Castle-an-Dinas Wolfram Mine (in 1957), East Pool and Agar Mine (in 1949) and Cligga Mine (in 1945); Great Rock Mine produced micaceous hematite until 1969. There are now active tin mines at Pendeen (Geevor Mine), near Camborne (South Crofty Mine and Wheal Pendarves) and at Baldhu (Wheal Jane), while Mount Wellington is likely to go into production.

The steady decline of the mining industry of south-west England from about 1880 to the slump of the 1920s was due primarily to the exploitation of colonial ore deposits. However, the present high market value of tin, caused by an impending world shortage, has revived interest in the Cornish deposits; intensive prospecting over selected areas during the past decade has resulted in renewed activity at a number of localities. Exploration at Levant has only recently been suspended.

Among the vast literature relating to the mining districts the works of Borlase, Pryce, Henwood, Phillips and Spargo in the late 18th and 19th centuries are invaluable, while the contributions of Collins, Davison and, more recently, Dines, J. H. Trounson and K. F. G. Hosking have added much to detailed knowledge of the mines and the processes of mineralization.

The stanniferous gravels mentioned above were probably deposited in and along valley bottoms as a result of the weathering of the cassiterite-bearing ore bodies and country rock under periglacial conditions during the Pleistocene Period. In places they are covered with a succession of river terrace deposits, and in the search for pebble-tin or stream-tin, as it was more generally known, the 'streamers' carefully followed the buried river channels; the worked-out areas are now represented by partly obscured trenches flanked by low dumps. The searches were mainly in the lower parts of the rivers, especially at Marazion, Carnon, Par and Pentewan, but some upland marshes were also worked, for example Goss Moor, west of Bodmin, and around the Birch Tor and Vitifer mines on Dartmoor. Small quantities of gold have been recovered from nearly all of these alluvial tin deposits and a nugget over 1 ounce in weight was discovered in the Carnon valley.

Over the centuries the introduction of improved separation techniques led to repeated reworking of these gravels, and eventually it became profitable to treat the effluent and slimes—bearing finely crushed cassiterite in suspension—from the treatment mills of working mines. Clearly an appreciable amount of cassiterite and other heavy minerals must have been swept out to sea and redeposited among sea-bed gravels and constantly shifting beach sands, and recently both St. Ives Bay and Mount's Bay have been dredge-sampled in an attempt to evaluate such deposits.

The source and areal distribution of the mineralization of the region is generally believed to be intimately related to the intrusion of the Armorican

granite, while later orogenic movements also produced primary mineralizations, as shown by age determinations of certain uranium minerals.
Both mining development and geophysical studies (gravity surveys) suggest
that belts of intense mineralization are related to ridges and 'cusps' in the
granite roof, many of which coincide with the 'emanative centres' of mineralization of Dines (1934). This relationship appears to be best displayed between
Land's End and the Bodmin Moor Granite; to the east Tertiary faulting has
tended to obscure this pattern.

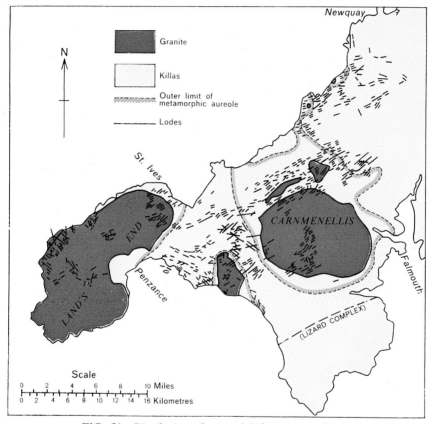

FIG. 21 *Distribution of mineral lodes in west Cornwall*
After 'The metalliferous mining region of south–west England', *Mem. Geol. Surv.*, 1956

The mineralizing fluids may have been a late-stage product of the consolidating granite or they may have risen through the slowly crystallizing granite
from great depth—possibly from the upper mantle of the earth's crust. The
former hypothesis is generally accepted for ore-genesis in south-west England;
during the cooling of the granite pockets of magmatic residuum and volatile-
rich fractions were probably trapped in the cusps until build-up of pressure
beneath the overlying strata allowed the mineralizing solutions to break free
and to pass upwards through fissures. The metallic (ore-forming) elements
were possibly largely derived from the melting of sediments and basic rocks

at depth which produced the granite batholith itself. Under favourable physical and chemical conditions the hydrothermal fluids deposited all or part of their metalliferous content within available structures in both the roof zone of the granite and the adjacent metamorphosed sediments (Figs. 22 and 23).

The intrusion of quartz-porphyry, aplite and thin pegmatite dykes is also associated with the late stages of granite emplacement and the dykes and mineralized structures tend to be parallel and to follow roughly the margins of the underlying granite ridges or even the batholith itself. West of the St. Austell Granite the general trend is north-east–south-west; to the east it is east–west. Among the exceptions are the crudely radial distribution of veins around the Land's End Granite, the structures around Perranporth and Gwinear Road, which are at a considerable distance from the nearest exposed granites, and also, in the latter area, the occurrence of two sets of dykes and veins with markedly differing trends. The main emplacement sequence was (1) granite, (2) pegmatites and porphyries and (3) mineral veins. Of the exceptions to this order Hosking (1952) recognized pegmatites formed before and during consolidation of the granite. At Castle-an-Dinas Wolfram Mine late granite apophyses penetrate quartz-wolfram veining. Collins (1912) noted porphyries cutting tin lodes and displacing cassiterite vein-swarms and Hosking (1966) recorded a porphyry cutting a feldspathic wolfram-arseno-pyrite vein.

The Lodes
'Lode' is a term probably derived from the verb 'lead' and applied by early miners to formations which might guide or lead to ore. In south-west England faulting has played a major part in governing the pattern of lode development and many hypothermal (high temperature) and mesothermal (intermediate temperature) veins can be directly related to normal faults. The occurrence of lodes in wrench faults and reverse faults is less pronounced. A general fracture pattern, which eventually produced the channels for the mineralizing fluids, probably developed during the initial period of folding and granite intrusion. Later relaxation of the regional pressures and the shrinkage caused by the cooling granite then reactivated the fractures, causing block movement and possibly gravity-slumping. Variations in the stresses from area to area established distinct local fracture patterns differing from the regional system; the fracture sequence at Geevor Mine has been described in detail by Garnett (1961).

Many of the tin lodes coincide in trend with one set of the major joints in the granite. Later fractures, roughly at right angles to the lodes and called cross-courses, generally follow another set of granite joints; they may carry mesothermal ore deposits such as lead and zinc but are commonly barren, and clay-filled cross-courses are known as fluccans. Lodes which diverge sharply in trend and dip from the general attitude of the lodes in any particular locality are called caunter lodes. The term referred originally to lodes striking east-south-east but is now so loosely applied that it may include branches of the normal-striking lodes. Most of the lodes dip at more than 70°, but a few dip at 45° or less, for example the Great Flat Lode south of Carn Brea (Fig. 22). In west Cornwall lodes dipping to the south commonly

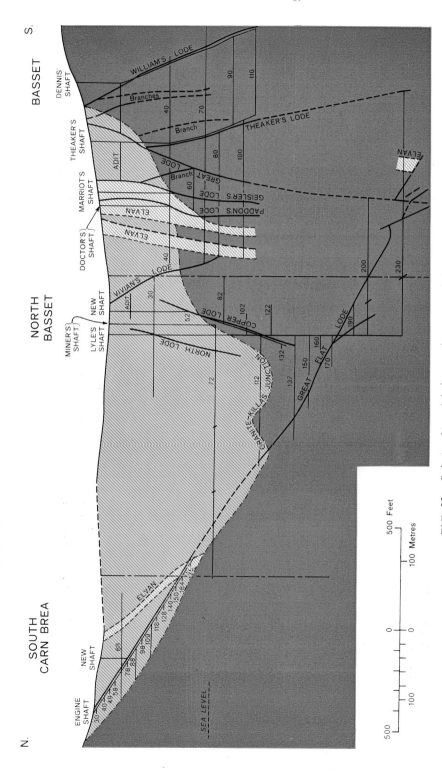

FIG. 22. *Relationship of lodes to the granite-killas junction*

After 'The metalliferous mining region of south-west England', *Mem. Geol. Surv.*, 1956

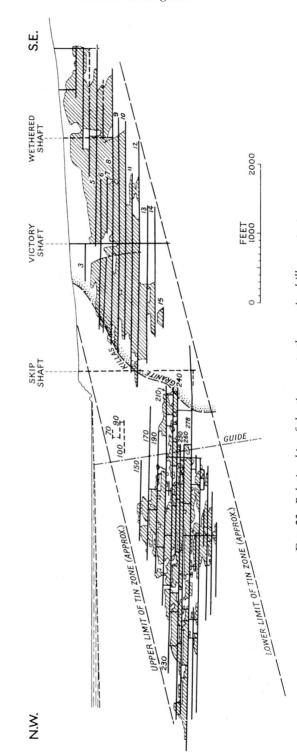

FIG. 23. *Relationship of the tin zone to the granite–killas contact*

Based on 'The metalliferous mining region of south-west England', *Mem. Geol. Surv.*, 1956

displace those dipping north and are generally richer in cassiterite but poorer in mixed sulphide and tungsten ores. In east Cornwall and Devon there is less difference between the two groups.

The lodes are not uniformly rich in metallic ores and their content varies considerably both laterally and in depth. In general they tend to be richest at changes of strike, intersections with other structures and also in the steeper sections of the lodes. To test the value of a lode samples are taken at closely spaced intervals and assayed, the results being expressed either as the ratio in pounds of the metallic substances to tons of rock waste taken from a measured width of the lode or as an average percentage metal content over the minimum stoping width of that particular lode.

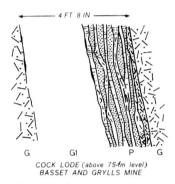

4 FT 8 IN

G   G1   P   G

COCK LODE (above 75-fm level)
BASSET AND GRYLLS MINE

G – Chloritized granite
G1 – Altered granite with some disseminated cassiterite
P – Blue tourmaline–quartz–peach with quartz strings and
    interstitial cassiterite. Soft band at hanging wall

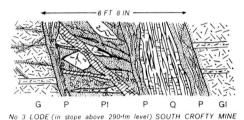

8 FT 8 IN

G   P   P1   P   Q   P   G1

No 3 LODE (in stope above 290-fm level) SOUTH CROFTY MINE

G – Granite with pink feldspars, thin tourmaline veins and
     horizontal veins of quartz and feldspar
G1 – Grey tourmaline–granite with horizontal quartz veins
P – Chlorite–peach with thin hematite strings and quartz veins
     with feldspar
P1 – Brecciated quartz and chlorite–peach lode material cut by
     diagonal tourmaline–peach band and tourmaline strings
Q – Banded quartz vein with thin iron-ore partings

3 FT

S   Q   S

WOLFRAM LODE (back of 3rd level)
CASTLE–AN–DINAS MINE

S – Soft white or buff shale, vertically bedded, passing into
    quartz–tourmaline–schist traversed by quartz veins near lode
    and locally impregnated with cassiterite at contact
Q – Milky white quartz with bunches of wolfram

2 FT 6 IN

G   G1   P   Q   I   Q   P   G1   G

NORTH LODE (at 11th level) GEEVOR MINE

G – Grey granite
G1 – Altered granite with red-stained feldspars, bearing some
     cassiterite
P – Dark tourmaline–cassiterite – peach
     iron-stained and carrying sulphides
Q – White vein-quartz with comb-structure
I – Siliceous brown hematite

FIG. 24. *Some variations of composite 'normal' lodes in Cornwall*
After 'The metalliferous mining region of south-west England', *Mem. Geol. Surv.*, 1956

The various types of Cornish lodes are as follows:

(*a*) *Mineralized faults.* The 'normal' lodes are mainly steeply dipping fault zones which have been repeatedly opened and mineralized; they may be infillings of clean-cut faults, mineralized fault-breccias or even mineralized shattered wall-rock on one or both sides of the main fracture (Fig. 24).

(b) *Replacement 'veins'*. Chloritization and silicification, locally accompanied by tourmalinization, are, in places, sufficiently intense to completely replace granite, aplite and porphyry dykes and occasionally adjacent slates. The replacement may be of the wall-rock flanking faults or it may be remote from obvious fissures. Where the metasomatizing fluids were rich in metallic elements the replacement bodies commonly form payable ore deposits.

(c) *Carbonas, pipes and floors*. Ore-bodies of this type are formed by rich local replacement of wall-rock (most commonly granite) by solutions migrating along minute fractures. Carbonas are best known in the St. Ives district (Providence and St. Ives Consols mines) where they comprise rather wide and near-vertical stanniferous bodies of highly altered granite. Pipes are similar but very narrow. Both types can usually be traced to narrow feeder channels and, in some mines, occur at the intersection of major structures of differing strike. Floors are sub-horizontal ore-bodies, usually arranged in tiers, produced by mineralization along flat-lying joints or faults connecting two vertical veins.

(d) *Stockworks and vein-swarms*. Vein-swarms are generally associated with the fracturing of granite cusps and usually occur within the granite (e.g. St. Michael's Mount, Kit Hill and Cligga Head), but in places they extend for a considerable distance into the overlying metamorphosed sediments as at Mulberry, near Bodmin. The swarms consist of numerous sub-parallel, steeply dipping, thin veinlets generally bordered by greisen and carrying wolfram, cassiterite and arsenopyrite; the strike direction of the swarms closely follows that of lodes flanking the cusps. Stockworks comprise an intricate network of veinlets and are generally found in the sedimentary strata above the cusps, as at Fatwork and Virtue Mine, although that at Hemerdon is in granite. Though greisenization did not occur at the former locality the veinlets bear a mineral assemblage similar to that of the vein-swarms. Smaller stockworks have developed in some cases by the mineralization of fractured dyke rocks or of the fissured hanging walls of lodes, the fissures having been produced by normal fault drag stresses.

Mineral Zones

In south-west England both the metallic and gangue minerals are distributed in a series of concentric belts or zones around the main emanative centres. The mineral zones can be related laterally and in depth to the thermal gradients which existed between the hot granite magma and the cool land surface; minerals crystallizing at high temperature (and under favourable pressure conditions) such as cassiterite, wolfram and tourmaline were deposited nearest to the granites and successive, cooler, outer zones formed suitable environments for copper, lead, zinc and finally iron mineralizations. Since the minerals in each zone crystallized under a particular set of temperature-pressure conditions the upper and lower limits of their occurrence in depth can be regarded as being representative of geo-isotherms (planes of uniform temperature) existing in the country rock at the time of mineral deposition. In general the mineral zones are inclined less steeply than the neighbouring granite–killas contact, and above the granite ridges and cusps they are more closely packed. Dines (1934; 1956) contended that successively higher mineral zones occupied greater lateral extents owing to

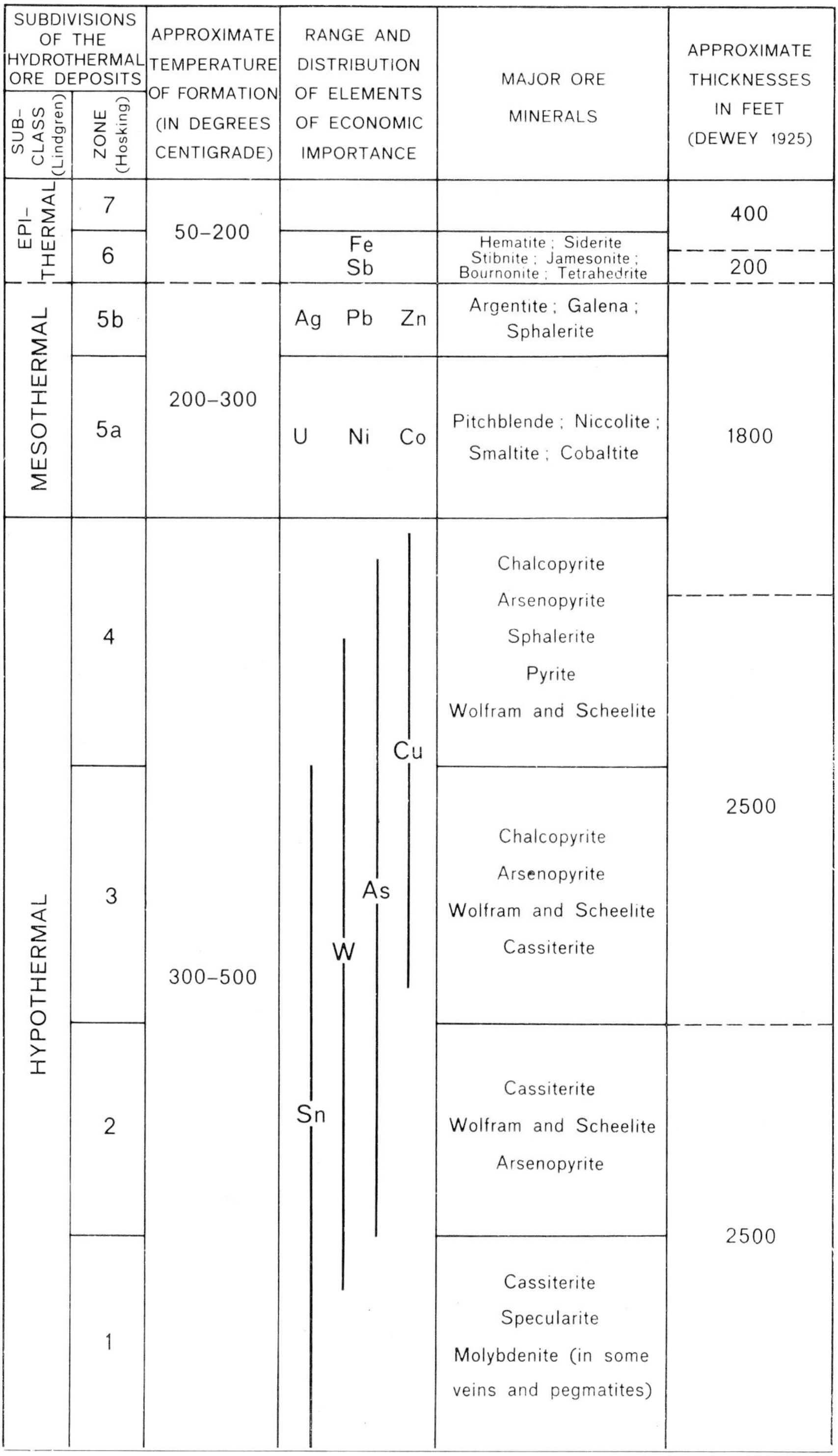

SUBDIVISIONS OF THE HYDROTHERMAL ORE DEPOSITS		APPROXIMATE TEMPERATURE OF FORMATION (IN DEGREES CENTIGRADE)	RANGE AND DISTRIBUTION OF ELEMENTS OF ECONOMIC IMPORTANCE	MAJOR ORE MINERALS	APPROXIMATE THICKNESSES IN FEET (DEWEY 1925)
SUB-CLASS (Lindgren)	ZONE (Hosking)				
EPI-THERMAL	7	50–200			400
	6		Fe Sb	Hematite ; Siderite Stibnite ; Jamesonite ; Bournonite ; Tetrahedrite	200
MESOTHERMAL	5b	200–300	Ag Pb Zn	Argentite ; Galena ; Sphalerite	1800
	5a		U Ni Co	Pitchblende ; Niccolite ; Smaltite ; Cobaltite	
HYPOTHERMAL	4	300–500	Cu As W Sn	Chalcopyrite Arsenopyrite Sphalerite Pyrite Wolfram and Scheelite	2500
	3			Chalcopyrite Arsenopyrite Wolfram and Scheelite Cassiterite	
	2			Cassiterite Wolfram and Scheelite Arsenopyrite	2500
	1			Cassiterite Specularite Molybdenite (in some veins and pegmatites)	

FIG. 25. *Schematic vertical relationships of the ore-mineral zones in south-west England*
Based on H. Dewey 1925 and K. F. G. Hosking 1966

the fact that the fracture pattern, initiated in and around the granite, became more open in the overlying country rock.

Changes in local conditions often resulted in the overlapping of mineral zones or even reversal of the order of deposition; notable exceptions to the general zonal arrangement include the occurrence of payable tin with sphalerite, chalcopyrite and pyrite at Wheal Jane, east of St. Day, situated beyond the metamorphic aureole of the Carnmenellis Granite, and the iron lodes worked at Ruby Mine and Treffry Consols within the St. Austell Granite. Such variations can tentatively be explained in terms of renewed hypothermal mineralization with further localized increases in temperature and pressure, changes in the metallic content of the mineralizing fluids or perhaps in the length of time between the opening of the fissures and actual mineralization. Generally, however, in any given lode, whether hypothermal or mesothermal, the minerals tend to 'young' upwards (Fig. 25).

Fig. 26 illustrates the zoning of the 'gangue' minerals which constitute the generally uneconomic and non-metallic rock waste associated with ore-bodies. In Cornwall the gangue species found in and around the hypothermal lodes are mainly adularia, fluorite and white mica (in greisen), tourmaline, chlorite and hematite. Quartz occurs as a gangue mineral throughout the whole temperature range and silicification is the major alteration process accompanying mesothermal mineralization. Baryte, dolomite, calcite and fluorite are characteristic gangue minerals in low-temperature veins.

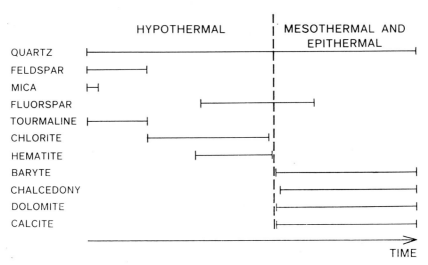

FIG. 26. *Generalized paragenesis of gangue minerals in south-west England*

Generally the pneumatolytic alteration processes affected lodes and wall-rock in both granite and killas country. In the latter greisenization was less common than sericitization and in some areas these were followed by the more widespread effects of tourmalinization and chloritization and later by kaolinization. Fine-grained aggregates of quartz, chlorite and tourmaline in the lodes are known to the Cornish miners as 'peach'; the chlorite-rich and

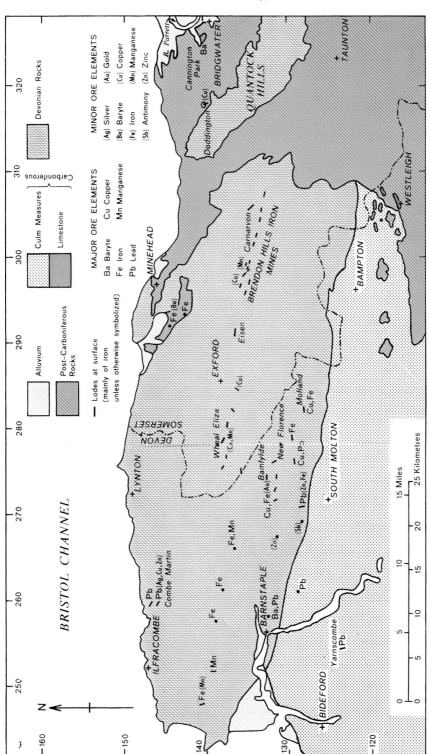

FIG. 27 *Distribution of metallic ores and baryte in north Devon and west Somerset*
Based on 'The metalliferous mining region of south-west England', *Mem. Geol. Surv.*, 1956

tourmaline-rich aggregates are referred to as 'green peach' and 'blue peach' respectively. These aggregates may contain payable quantities of tin ore.

The basic principles of the mineral zones were proved in individual mines as they worked downwards; for example, Dolcoath Mine yielded a little zinc with copper from the upper levels, then copper alone down to 170 fathoms, mixed copper and tin between 170 and 190 fathoms, and only tin below to the bottom of the mine at 550 fathoms. At the nearby East Pool Mine the copper zone extended to 140 fathoms, the wolfram zone thence to 200 fathoms, and below that tin. In the St. Austell–Par district several mines illustrated the upward passage from the copper zone directly into a zone of iron carbonate (siderite), but elsewhere the normal passage from copper zone to lead zone occurs. The maximum depth to which lead has been worked in the region is 300 fathoms at Wheal Mary Ann, near Liskeard, but generally the lead zone averages only 140 fathoms (Dewey 1925). Farther away from the granites siderite (and its oxidation product, brown hematite) commonly occurs in the lodes of Exmoor, around south-east Dartmoor and also in the Great Perran Lode. The latter, however, is an east–west fracture system which has been mineralized more intensely at intersections with later north–south lead-zinc-siderite structures but also carries traces of primary chalcopyrite and pyrite at depth; the mineral assemblage suggests that the lode resulted from hydro-thermal mineralization though its strike differs from most of the Cornish iron lodes.

Other mesothermal deposits remote from the granite outcrops include the antimony ores to the north-west of Wadebridge, the lead, silver-lead and baryte veins of the Barnstaple–Combe Martin area, and the baryte mineraliza-tion of Cannington Park and the northern Quantock Hills (Fig. 27). At Doddington, on the eastern fringe of the Quantocks, chalcopyrite and malachite were worked during the early part of the 19th century.

Secondary enrichment

Secondary alteration of the lode minerals above the water table is generally complete throughout the region. With a few exceptions the zones of leaching and enrichment have long since been eroded or worked away; these changes, which were effected by downward percolating rain-water carrying carbon dioxide, locally extended to 150 fathoms below the surface. As a result of this solution the upper parts of the lodes commonly consist essentially of gangue minerals with iron and manganese oxides which form the 'gossan' or 'iron hat' of the miner (Fig. 28). The gossans over some galena veins have pre-served traces of native silver and it is believed that much of the gold obtained from the alluvial deposits was derived from the erosion of gossans overlying copper lodes. Beneath the gossan the effects of leaching may have resulted in a barren zone from which the soluble minerals have been removed—in the case of primary chalcopyrite the copper is generally transported as a soluble sulphate—only to be redeposited immediately above the water table as black and red oxides, and blue and green carbonates of copper. Where the copper-bearing solutions permeated the lode below the water-table (where oxygen is excluded) the copper may have been precipitated by reaction with the original copper sulphides to form bornite and chalcocite. These minerals are characteristic of the zone of 'secondary sulphide enrichment' in which the ore yields an increased percentage of copper—the original chalcopyrite contains 34·5 per cent copper, bornite 63·3 per cent and chalcocite 79·8 per cent. Typical minerals in the enriched oxidized zones of lead and zinc veins are the carbonates cerussite and smithsonite.

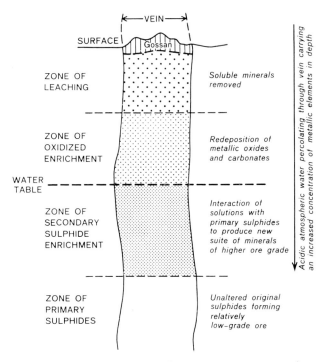

FIG. 28. *Diagrammatic section illustrating supergene enrichment*

Lithological control of mineralization

Many examples have been described of changes in the mineralogical characters of lodes as they traverse different rock types. Henwood (1843) gave several instances of lodes bearing copper in sediments and greenstones, and tin in granite, but there are too many exceptions (e.g. at Wheal Vor productive tin ore occurred in the slates but the lode became barren in the granite) for this to be accepted as a general rule in the Cornish mining area. Lead lodes, however, are absent in the granites, rare in the metamorphic aureoles, and within the sediments have been proved in several mines to have been preferentially deposited in the shaly rocks, the interbedded siliceous or more sandy beds being barren. This feature is well exhibited in the silver-lead lodes of the Chiverton Mines, near Perranzabuloe (Fig. 29).

Within the metamorphic aureole of the Dartmoor Granite both tin and copper mineralization have been recorded from trials and mines worked along the crop of the Meldon Chert Formation and the adjoining Meldon Shale and Quartzite Formation, particularly between Fanny Mine, near Bridestowe, in the west and Ramsley Mine at Sticklepath in the east. The tin mineralization appears to be concentrated within the Chert Formation in skarns and narrow bands of wollastonite-hornfels which contain stanniferous garnets and stanniferous sphene (malayaite) respectively. Chalcopyrite, together with arsenopyrite, löllingite, pyrite and pyrrhotite, is patchily distributed in restricted horizons of calc-silicate hornfelses rich in garnet, axinite and diopside. The ore horizons or beds are generally softer than the enveloping hornfelses; at Belstone Mine four such beds were exploited and at Ramsley Mine three. At the same stratigraphic level in south-east Dartmoor magnetite deposits were worked at Haytor Mine; the ore consists of fine aggregates of magnetite and hornblende in beds up to 14 ft thick interbedded with siliceous shales and sandstones. Three beds were developed and immediately above the lowest a narrow transgressive fine-grained granite sill was proved; Dines (1956) stated that these deposits are generally held to be due to emanations from the granite but suggested that the emanations might have been derived from the pre-granite epidiorites of the district.

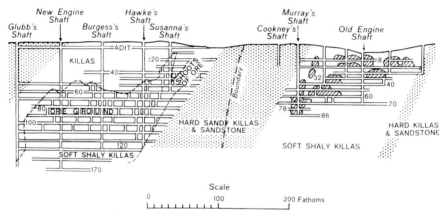

FIG. 29. *Section illustrating lithological control of lead mineralization in the Chiverton Mines*
After 'The metalliferous mining region of south-west England', *Mem. Geol. Surv.*, 1956

Magnetite, probably derived from basic igneous rocks, occurs as segregated lenticular masses within the country rock at Botallack, Trebarwith and Tintagel. Manganese deposits in hard cherty beds in the Milton Abbot area, west of Dartmoor, are generally associated with basic volcanic rocks and indicate preferential mineralization in the Lower Carboniferous strata.

Iron ores
In Cornwall, particularly between St. Austell and Wadebridge, most of the iron lodes trend north–south; those within the granite, such as the Ruby Lode (north-east of St. Austell), and also the Lanjew, Wheal James–Retire, Nanstallon and Restormel group of lodes lying to the north and east of the granite consist mainly of red and brown hematites. The micaceous hematite

(Photo: Gibson, Penzance)   *(A 8080)*

A.   Botallack Mine in operation, about 1870

Plate 11   *(For full explanation see p. xi)*

B.   Cligga Head Quarry, greisenized granite

(A 8124)

(*Photo: The Old Delabole Slate Co. Ltd.*)

A. Old Delabole Quarry

Plate 12   (*For full explanation see p. xi*)

B. China clay workings, St. Austell

(*A 9801*)

lodes at Great Rock Mine and in the surrounding neighbourhood of east Dartmoor trend east-north-east; the granite wall-rock is commonly chloritized, and beyond the lodes it is traversed by small tourmaline veins. In north Devon and west Somerset the iron lodes are primarily of siderite, which near the surface has been weathered to goethitic brown hematites of generally low phosphorus content. The lodes follow the regional west-north-west trend of the Devonian strata and occupy a belt of country a mile or so wide in the Brendon Hills district, increasing to several miles in central Exmoor; the ore-bodies appear to have been controlled by the southerly-dipping bedding and cleavage of the surrounding slates and sandstones, the occurrences of ore commonly being sporadic and connected by clay-filled fissures or narrow quartz strings. Mines in an east–west belt to the north of North Molton (particularly Bampfylde and Molland mines) yielded both iron and copper ores, the latter chiefly chalcopyrite.

Uranium ores

Traces of the primary uranium minerals pitchblende, uraninite and coffinite and secondary alteration species such as autunite and torbernite are found in most parts of the region. The uranium mineralizations are associated with the major granite bosses and are generally found in crosscourses accompanying mesothermal minerals such as those of iron, cobalt, nickel and bismuth (South Terras, St. Austell Consols, Wheal Bray and King's Wood mines), more rarely with lead-zinc ores (Wheal Owles) and also, in the case of Geevor and South Crofty mines, in association with hypothermal tin ore. The secondary minerals commonly line joints in the granite. There have been several periods of uranium mineralization. Wheal Trenwith at St. Ives and South Terras Mine near St. Stephen were the main producers of uranium ore.

Bulk minerals and building materials

Clay

The name kaolin comes from that of the mountain Kauling, in China, near where the material was discovered. China clay (p. 49) is the most valuable of our raw material exports and the reserves in south-west England, largely near St. Austell and Lee Moor, are great (Fig. 30); production at Lee Moor exceeds 500 000 tons a year. Perhaps the most remote of the clay works was that at Redlake, 1400 ft above sea level on southern Dartmoor, which closed in 1932. Large-scale exploitation began in the 18th century and expansion during the latter part of the 19th afforded employment to the increasing numbers of tin miners thrown out of work. In recent years the industry has become intensively mechanized. High-pressure jets of water strike the working face of china clay pits with such force that sizable boulders are thrown high into the air (Plate 12B). Quartz grains and then mica are removed from the resultant slurry by sedimentation in channels and settling tanks. The sandy waste is now used in bricks and concrete products and the finer 'micas' have been marketed as low grade china clay and used in oil refining as an absorbent and detergent.

Much of the clay is used in paper manufacture, some goes to the potteries, and among the varied minor end products are face powder and indigestion remedies. Between about 1769 and 1836 Coade stone was made by heating china clay with quartz, flint and glass; it became popular for both interior and exterior decoration, being hard and weather proof, but the vogue died out.

Pottery clay is dug from mines and open workings (Plate 10A) in the Oligocene deposits (pp. 77–8) at Bovey Tracey and Petrockstow—in the former area since the 1730s. Initially it was cut into rough cubes each weighing about 36 lb, and from their unlikely local name of balls was derived the term ball clay to describe the raw material. The clay is used for drain pipes, building bricks, acid-proof bricks, firebricks, chimney pots and tiles; the manufacture of high quality white earthenware was formerly a flourishing local industry and large quantities of clay are now sent to the potteries. The plasticity of ball clay gives a strong bond to the pottery but the fired ware is commonly cream coloured; a mixture of ball clay and china clay may be used to combine strength with whiteness.

Many cottages and (mainly small) houses, farm buildings and garden walls are made of cob. A stone foundation is usual and on it walls are built up by successive layers of a sticky mixture of clay, water, chopped straw and occasionally animal hair. Cloam ovens, of earthenware, may be incorporated in the cob walls. The clay has generally been dug as near as possible to the building site; it is commonly yellowish brown stony Head but is typically red on Permian and Triassic ground. Given overhead protection cob walls will stand for many years; the probably 13th century coin found in the wall of a Dartmoor cottage is not conclusive evidence of age, but cob was used in the 15th century Trelawne Manor, Pelynt. Present-day use is confined to the repair of existing cob structures.

There are few large brickworks outside the small areas of Oligocene clay. The Western Counties Brick Company has a number of pits in south Devon and Cornwall between Torquay and Millbrook, the North Cornwall Brick Company has a works at Bridgerule, near Holsworthy, and there are one or two small active pits in the Barnstaple area. Numerous derelict flooded clay-pits mark the sites of small brickworks which were opened to meet local needs, commonly worked clay of variable or poor quality, and closed down when the demands of their immediate neighbourhoods ceased. Medieval tiles may be seen in the church floors at Launcells and Abbots Bickington.

Stone

A variety of rocks has been quarried for use as railway ballast, roadstone and concrete aggregate—granite, greenstone, dyke rocks, slate, shale, sandstone, limestone and hornfels. About 286 000 cu yd of crushed stone were taken from the huge British Rail Quarry at Meldon in 1964. The most westerly railway in England is that bringing greenstone from Penlee Quarry to the harbour at Newlyn.

Limestone and chalk still provide some agricultural lime but most lime-kilns, which commonly date from around 1800, are now disused. Beer Freestone (pp. 73–4) has been cut from the Chalk since medieval times and used as a building stone in several cathedrals and smaller churches.

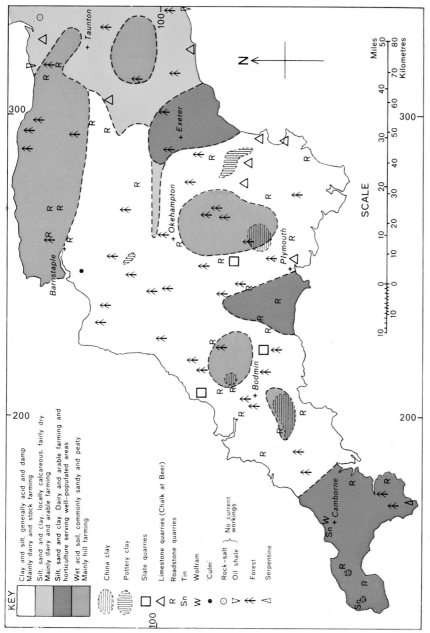

FIG. 30. *Soils, land use and economic geology*

KEY

	Clay and silt, generally acid and damp Mainly dairy and stock farming
	Silt, sand and clay, locally calcareous, fairly dry Mainly dairy and arable farming
	Silt, sand and clay. Dairy and arable farming and horticulture serving well-populated areas
	Wet acid soil, commonly sandy and peaty Mainly hill farming

Symbol	Description
	China clay
	Pottery clay
□	Slate quarries
△	Limestone quarries (Chalk at Beer)
R	Roadstone quarries
Sn	Tin
W	Wolfram
•	'Culm'
⊙	Rock-salt } No current
▽	Oil shale } workings
⇇	Forest
◁	Serpentine

Quarrying of slate too dates from the Middle Ages but its heyday as a roofing medium was the 19th century when it replaced thatch and 'stone slates'. The huge hole at Delabole (Plate 12A) is probably the biggest slate quarry in the world but new outlets have had to be found for the stone: floor slabs, crazy paving, tombstones, memorial tablets, fireplace surrounds and hearths are produced in addition to roofing slate, and the rock is ground for use in concrete products and as a filler in certain chemical products.

A 12th century doorway in Landewednack church evidences the working of serpentine at that time. It is still quarried on the Lizard and the variegated red, green, blue and black rock is turned on lathes and made into knick-knacks for sale to tourists. Working of Polyphant Stone, a serpentinized picrite, dates from the 11th century; it has been used in Launceston Castle, Truro cathedral and Exeter cathedral.

Until about 1800 granite for building was taken from the vast amounts of surface debris. It was used for cottage and hut foundations, for larger houses, churches, ancient tombs, guide posts, crosses, gate posts, bridges, stiles, stepping stones, millstones, rollers, presses, farm troughs and paving stones. The famous clapper bridge at Postbridge probably dates from the 13th century. Rounded knocking stones have served for centuries for pounding cereals for the farmer and green gorse for his cattle, and the early tin streamers used granite for their blowing houses and ingot moulds. Since the 15th century granite has been used in buildings both as roughly hewn blocks and cut into ashlar. In recent years quarrying has declined although there is a small demand for building blocks of ground 'reconstituted' granite where planning regulations demand the use of stone.

One of the more interesting uses of granite was in the Haytor horse tramway. Quarries here supplied stone for use in London in the National Gallery, the old Post Office, the British Museum and London Bridge, and were connected by a granite tramway of 4 ft 3 in gauge to the Stover Canal at Teigngrace. Rails of granite 4 to 8 ft long, about 15 inches wide and with 2- to 3-inch inside flanges may still be seen although the tramway, which climbed 1200 ft in 6 miles, was closed in 1858. A mineral line from the Kilmar quarries on the south-east side of Bodmin Moor closed in 1916. Granite from Princetown, together with slate and elvan from Cann Quarry, was the main traffic of the old Plymouth and Dartmoor Railway and granite from Kit Hill of the upper stretches of the East Cornwall Mineral Railway.

The Meldon Aplite was used intermittently between the 1880s and about 1920 for the local manufacture of glass.

Sand and Gravel

Little extensive digging of sand and gravel for building purposes has taken place in south-west England, although a few commercial pits exist and countless small diggings occur where river terraces, Upper Greensand and Tertiary gravels have been exploited locally. Thick sand and gravel terrace deposits underlie Taw Marsh and have been tapped for water. Moulding sand was formerly dug at St. Erth.

Shelly beach sand has been much used as a fertiliser in several areas. The St. Columb Canal carried such sand from Trenance Point to Lower St. Columb and imports of the sand at Newton Abbot are recorded. Off Looe, sand was dredged from the sea bed by means of iron-braced canvas bags and carried in barges to Sandplace and thence initially by pack-horse but later by canal and railway to Moorswater, near Liskeard. Sea sand was also carried inland from Wadebridge to Bodmin on Cornwall's first locomotive line. It has been dug from the Bude area for very many years and was the main reason for the construction of the Bude Canal, an ambitious design using inclined planes instead of locks, which was opened in 1825, ran from Bude to near Holsworthy with a branch to Druxton and was fed from a reservoir at Alfardisworthy.

Peat

Only on high Dartmoor are there wide areas of blanket peat, the thickness of which ranges up to 20 ft or more. The Cornish granites are at too low an altitude for its widespread development, although patches occur above 1000 ft O.D. on Bodmin Moor—as on Brockabarrow Common, where peat has been cut. Peaty Head has accumulated in the upper stretches of many moorland valleys.

The early tin streamers used peat both as a fuel and as a separating surface which trapped the ore when washings were run over it, and it has been cut at many other places, usually in random fashion on a very small scale for domestic fuel. In 1878 the West of England Compressed Peat Company opened its Rattle Brook works below Kitty Tor on the north-west slopes of Dartmoor. The following year a mineral railway was built connecting the works with Bridestowe but operations continued for only a few years. Further uneconomic schemes, in 1917–19 and in 1943, entailed the establishment of chemical works at the site, but these too failed. The railway was ripped up in 1931–2 but peat has since been occasionally dug, and carried by lorries; a quantity was delivered at Bridestowe Station for transport to Guernsey in the mid-1950s. The Zeal Tor peat tramway was opened in 1847 between the edge of the moor and Shipley, near South Brent, and later carried clay.

Probably there is no economic future for the large peat reserves of northern Dartmoor but even if exploitation was proposed it must be remembered that the peat cover acts as a giant sponge, retaining water long enough to allow much percolation into underlying weathered granite. Removal of the peat would result in even more rapid run-off, an effect which could be countered only by extensive afforestation.

Miscellaneous Products

Rock-salt and Gypsum

Salt was discovered in boring for coal at Puriton, Somerset, in 1910, where saliferous strata within the Keuper Marl extended from $646\frac{1}{2}$ ft depth to $719\frac{1}{3}$ ft. The beds comprised marl, sandy marl and some sandstone with rock-salt and gypsum, and the rock-salt was recovered in small pieces and probably much was lost by solution during drilling. Two more bores were sunk and a salt works operated there for a few years. The percentage composition of the rock-salt was sodium chloride 97·47, calcium sulphate 2·13, calcium chloride 0·15, magnesium chloride 0·25, and a 5-month test of the brine showed an average yield of 1 lb 13 oz salt per gallon.

In the 19th century gypsum was collected from the Keuper Marl at the coast near Watchet but no mining has been attempted. Gypsum has, however, been worked between Branscombe and Weston Mouth.

Coal, Pigment and Oil

Lignite in seams up to 16 ft thick has been dug from the Oligocene beds of Bovey Tracey; it is said to have emitted an offensive smell when used as a domestic fuel. Thin discontinuous beds of 'culm' have long been known between Bideford Bay and Chittlehampton and their outcrops are marked by shallow pits. Small-scale mining for fuel was attempted during the last century just east of Bideford, working a seam said to have ranged from 6 inches to 14 ft in thickness. The material was worked until 1969 as a pigment ('Bideford Black'). Traces of coal are common elsewhere in the Carboniferous rocks but there is no prospect of development.

Ochre, an earthy hydrated iron oxide formed by the decomposition of pyrite, has been recovered from some twenty mines; when gritty particles are washed out the resultant fine yellow, red or brown paste is suitable for pigments. Dark brown ochre is known commercially as umber and has been produced at nine mines. However, true umber is a mixture of hydrated oxides of iron and manganese with some silica and alumina; it may result from the decomposition of limestone, as at the old Devon and Cornwall Umber Works at Ashburton.

Oil shale was discovered about 1900 in the Lower Lias of the coast at Kilve. Attempts to extract the oil commercially were frustrated by the high sulphur content and the workings were abandoned. Some of the Liassic shales near Lyme Regis are so highly bituminous that decomposition of the abundant pyrite within them has occasionally generated enough heat to ignite them. Thus smoke and flame were noted on the cliffs in August 1751, and in January 1908 so much smoke rose from a slipped mass of shale that the local people feared a volcanic eruption.

Soils

Large areas of south-west England are covered by poor soils unsuitable for arable farming (Fig. 30). The country rock is generally covered by locally-derived Head whose upper few inches have broken down into soil. A layer of strong iron-pan, locally known as 'black ram', is commonly present about 1 or 2 ft down and may interfere with already reluctant drainage.

Sandstones of Devonian and Carboniferous age break down into local sandy soils. Devonian shales and slates yield a fairly free-draining brown earth of silty and clayey loam and it is curious that except in markedly low-lying areas the underlying rock seems not to impede drainage as much as do the Carboniferous shales. Flat-lying areas of Carboniferous shales at both low and high levels are covered by wet clayey soils bearing rough rush-covered pasture. Extensive areas of this sort have been developed as forestry plantations. Such soils are extremely difficult to cultivate and although many have been brought into cereal production during wartime this requires heavy capital outlay such as few individual farmers can now contemplate. The increase in dairy farming on these poor clayey soils of Devon and Cornwall since the second world war is probably greater than in any other region; it has been possible because of a government price policy of subsidising milk transport in remote areas.

Friable sandy soils on the lower areas of granite and around the margins of the bosses have been tilled locally but the high moors of south-west England are covered by thin impoverished peaty soil which affords some rough grazing in summer and will support coniferous plantations except on the highest ground and on the blanket peat bog. Many moorland farms have ash-houses, small circular granite huts in which ash from the burning of wood, peat, gorse and heather was in the past stored for use as fertilizer.

Small strips of rich alluvial soil occur in the lower reaches of certain valleys but only in the eastern part of the region are there extensive areas of arable land. These are the 'Red Devon' soils developed on breccias, sandstones and marls of Permian age. Triassic pebble beds yield a sterile soil but rich grass may be grown on the heavy soil of the Keuper Marl. Soils on the Upper Greensand, particularly those on the Foxmould, are fertile and easily worked.

Water Supply

South-west England is an area of moderate or high rainfall and low population density and it has little heavy industry. Hence there are adequate water resources to meet the increasing needs of population, agriculture and such industrial development as seems probable at present. However, much of the area is underlain by Devonian and Carboniferous slates and shales of low porosity and permeability. Their generally small yields to wells and boreholes depend largely on the presence of fissures, which are best developed where the rocks have been baked by granite or affected by faulting. Fissure systems are rarely well interconnected, however, and supplies commonly diminish or fail in dry summers. Contamination and resultant bacterial pollution are common. Small quantities of hard water have been obtained from Devonian limestones near Newton Abbot, Torquay and Plymouth. Some water occurs in fissures in the granites but the permeability of fresh granite is less than that of the slates and shales. Weathering of the granite may extend to a depth of 50 ft or more, when good supplies of soft acid water can be obtained; conditions are even more favourable where the weathered rock is overlain by drift. Breccias and sandstones of Permian and Triassic age are water bearing, and are being developed for municipal supplies in the Otter Valley, but large areas of eastern Devon and western Somerset are underlain by thick impermeable Keuper Marl and Lower Lias clay from which no supplies are obtainable. The overlying Upper Greensand and Chalk are both water bearing. Where they occur together they are in hydrological continuity and yield hard water, but where the Greensand lacks a chalk cover, as in the Blackdown Hills, the ground water is soft.

Drifts yield small local supplies, commonly subject to contamination. Nine shallow wells have been sunk in thick terrace deposits of gravel, sand and silt overlying granite below Taw Marsh; total abstraction is about $1\frac{1}{4}$ million gallons a day and the water is soft and acid.

Water authorities of west Cornwall, east Cornwall, north Devon, south-west Devon and west Somerset draw most of their supplies from surface water—mainly from reservoirs and river intakes. In east Devon surface water is abstracted from rivers and springs and ground water from bore-holes. All large centres of population except Exeter depend almost wholly on river intakes and reservoirs, although some supplies in mining areas are drawn from old shafts and adits. Of local interest is Tamar Lake, which now supplies Bude and Stratton but was constructed as Alfardisworthy Reservoir to feed the Bude Canal. Tavistock Canal connected the town with the River Tamar; it closed in the late 19th century but was cleaned out in 1933–4 when the water was led to a small reservoir and thence piped to a hydroelectric plant at Morwellham. Plymouth now draws water from Burrator Reservoir but was noteworthy in the past for a 24-mile leat constructed in 1585 to bring water from Dartmoor.

Thus most of south-west England, and in particular that part west of Exeter and Tiverton, depends on surface water for the majority of large supplies. Small reservoirs tapped by gravity feeds can store water on high ground such as Dartmoor and Bodmin Moor; several have already been constructed and many other possible sites exist. However, there may be both practical and aesthetic objections to this.

The need for co-ordinated regional, rather than piecemeal, development of all the available water resources is paramount. The long-term answer, if we exclude desalination, may be large-capacity reservoirs, or possibly barrages across some of the many estuaries. Pumped-storage schemes may be appropriate for some areas, with fresh water being drawn from the estuaries and stored in reservoirs inland; such reservoirs could be multi-purpose and used for power generation, flood control and recreation as well as for water supplies for people and stock.

Potential water resources are such that the time may come when water is pumped from the west country to supply the drier, more populous, areas farther east, either directly by pipeline or indirectly by supplementing river flow.

II. Structure

The general synclinorial structure of south-west England, with Carboniferous rocks flanked to north and south by Devonian strata, was recognized by De la Beche (1839, pl. 1). Subsequent work has traced much of the intricate small-scale structural detail within this trough, so that a broad picture can be painted; but more remains to be done, especially in south-west Cornwall. Many workers have contributed to current views, but foremost among those who have attempted syntheses of the structure of the peninsula is Dearman (1964b; 1969; 1970; 1971). Both he and Simpson (1969; 1970; 1971) have divided the peninsula into tectonic zones, as have Sanderson and Dearman (1973), but most of their lines are necessarily arbitrary and most of the changes in fold style gradual (Figs. 31 and 32).

Around the end of the Silurian Period the Caledonian earth movements reached their peak. Their effects in Britain, most pronounced in northern Scotland and diminishing southwards, are mainly shown as a north-easterly structural trend. Such structures have been encroached upon and overprinted by later easterly-trending folds of the Armorican orogeny.

A Caledonoid zone in south-west Cornwall is truncated roughly along a line from Perranporth through Pentewan and Salcombe (the southern edge of the Lower Devonian of Fig. 1) by the undoubtedly Armorican zone which covers the rest of the peninsula. However, it is important to realise that use of the term Caledonoid in south-west Cornwall implies only a north-easterly trend, and does not necessarily relate the structures to the Caledonian fold belt. Such radiometric datings as are available (Dodson and Rex 1971) point to a Devonian metamorphic event, between the Caledonian and Armorican orogenies, but the evidence is inadequate to support firm conclusions. Radiometric dating of slates farther north (Fig. 33) gives younger ages in the north of the peninsula than in the south, but it must be borne in mind that these figures may relate to uplift and cooling rather than to metamorphism.

The Armorican orogeny, of late Carboniferous and early Permian times, was characterized by north–south compressive forces which have produced an east–west structural trend. It is folding of this age which is dominant over most of south-west England. Towards the end of the period of compression the Cornubian granite magmas, rising from great depths, approached perhaps to within some hundreds of feet of the surface. Strata were arched up over a number of cupolas and in places, as above the Bodmin Moor Granite, slices of pre-folded rock slid radially outwards from the rise. Some of this low-angle faulting took place by reactivation along earlier thrust planes. As the granites crystallised, and pressure relaxed, easterly-trending normal faults developed. North-westerly wrench faults may have been initiated by earlier compressive forces, but many were reactivated in Tertiary times and may still move to the extent of causing minor earth tremors.

In general the folds trend easterly, although Dearman (1964b) has suggested modification by Tertiary wrench faults. Within the great south-west England synclinorium major folds may be traced from the pattern of mapped outcrop, but the mass of field evidence upon which a structural synthesis must be based comprises the smaller folds displayed so abundantly on the coast and locally inland. These relatively minor structures are presumed to reflect the major pattern. They show (Fig. 32) that open upright folds north of Bude, in the centre of the Carboniferous basin, fan out northwards into northward-overturned folds and southwards into southward-overturned folds. The latter become recumbent and isoclinal towards Padstow, south of which the primary recumbent folds face north. A second phase of folding is common, and a third occurs in places. Tectonic events migrated northwards and it is probable that second-phase folds in the south are contemporary with the first-phase deformation in the north. Similarly, raised rocks lying in north-facing folds within a nappe-like structure south of Padstow probably provided a source of sediment for some upper Carboniferous deposition farther north. The pattern of primary folding may best be described by tracing the change in attitudes from north Devon to south Cornwall (Fig. 32).

In north Devon the Devonian strata form a large anticline whose axis trends east-south-eastwards through Lynton. South of that axis progressively younger formations dip southwards without large-scale repetition. Fold attitudes change from northward-overturned on the north coast to upright in the Pilton Beds of the Taw–Torridge estuary, and fold axes are sub-horizontal trending between east and east-north-east.

Simpson (1971) noted that the northward overturning was best illustrated by the change in attitude of the slaty (axial planar) cleavage within the more argillaceous formations, from 30°S. at Lynmouth to near vertical at Morte Point. In striking contrast is the competence of the Hangman Grits, the Pickwell Down Beds and the Baggy Beds, which are generally but slightly folded except in a regional sense. This contrast led Holwill and others (1969) to contemplate differential movement between arenaceous formations, with the Hangman Grits and Pickwell Down Beds acting as a couple on the Ilfracombe Beds and Morte Slates between. However, neither here nor in the comparable Pickwell Down Beds–Upcott Beds–Baggy Beds sequence farther south is there evidence of any major strike thrusts. Many of the formational boundaries are gradational, and it appears that fold movements were accommodated mainly by crumpling within the argillaceous strata. Widespread minor slipping occurred on bedding planes but there was no great interformational thrusting.

Near-upright folds characterize the Pilton Beds, the overlying cherts of the Lower Carboniferous between Barnstaple and South Molton, and the adjoining Upper Carboniferous strata (Fig. 13). There is no evidence of the extensive thrusting postulated by Prentice (1960a) and Reading (1965). Freshney and Taylor (1971; 1973), in establishing a stratigraphic succession for the coastline between Hartland Point and Bude, have confirmed the presence of folds overturned slightly to both north and south. Upright folds south of Bude fan over into south-facing recumbent folds farther south, beyond Wanson Mouth, perhaps marking the passage from the upper (normal) limb of a major fold into the overturned limb.

The large low-angle Rusey Fault marks a major tectonic break. To the south of a line from Rusey Beach through Tremaine, deformation is increased. The early recumbent folds are partly obscured by second-phase zigzag folding,

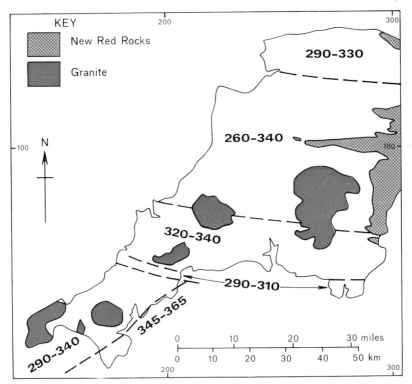

FIG. 33. *Potassium–argon ages of slates in millions of years*
Based on data from Dodson and Rex 1971

and there is extensive low-angle thrusting. Freshney and others (Tintagel–Bude Memoir, 1972) considered that these rocks may have been brought up from deeper tectonic levels, perhaps by faulting. The low-angle faults of the area appear to group round the Bodmin Moor Granite, whence they dip north and west delineating slices of strata which have been shed radially from above the rising magma. Effects of intrusion are thought also to include the development of the late zigzag folds cascading away from the centre of the granite.

Intense deformation around Boscastle and Tintagel extends inland towards Lydford, affecting topmost Devonian, Lower Carboniferous and basal Crackington Formation strata. Fold axial trends range from north-north-west to east-north-east, and facing directions from west-south-west through south-south-east to east-north-east. Overthrusting from the south of one slice of rocks by another has resulted in Lower Carboniferous strata being inserted into Upper Devonian as far south as Tintagel and Tregardock. The pattern is matched on the east side of Dartmoor, where several Palaeozoic successions (Table 2) have been identified in thrust sheets that represent progressively higher structural levels towards the south. Locally throughout this zone of extreme deformation it appears that recumbently folded strata have overridden upright folds farther north.

Folding to the south of this belt resembles that to the north, recumbent and south-facing with axes trending around east-north-east, and this pattern extends to a line roughly from Padstow through the southern ends of the Bodmin Moor and Dartmoor granites, a line along which occurs a major confrontation of south-facing folds to the north and north-facing folds to the south within the St. Minver synclinorium. Possibly the two opposed sets of folds are separated by a thrust or fault. It seems inconceivable that southerly and northerly movements could have occurred at the same time. Roberts and Sanderson (1971) have accounted for this contrast in terms of two phases of folding, the earlier effective in the south, overriding northwards and dying out northwards, the later effective in the north, overriding southwards and dying out southwards but producing some refolding of the earlier north-facing folds.

Whatever the explanation, north-facing folds predominate from Padstow to southern Cornwall. Immediately south of Padstow they are recumbent, trending east-north-east, but south-dipping fold axial planes around New-quay, and in a belt which extends eastwards to include most of the coast as far as Dartmouth, steepen southwards. The intensity of second-phase folding increases in the same direction. A major anticlinal axis, displaced locally by faults, runs eastwards within Lower Devonian rocks from Trenance, north of Newquay, to Dartmouth.

Sanderson (1971) suggested that the northern boundary of the Gramscatho Beds at Perranporth, and perhaps all along the Perranporth–Pentewan line, was probably a slide on which major north-facing recumbent folds moved north-north-westwards. The Mylor Beds and Gramscatho Beds between Perranporth and Penzance show recumbent first-phase folds trending east-north-east and facing north or north-west, second-phase upright folds and third-phase flat-lying folds. Exposures in the neighbourhood of Porthleven show primary folds whose axial planes dip around south-east. A similar pattern persists north-eastwards to Truro, but in areas of intense deformation north of the Lizard and Dodman Point some oblique folds plunge south-east. These minor folds lie within a major anticline trending east-north-east from north of Marazion through Truro.

The metamorphic rocks of the Lizard, Dodman Point and Start Point may represent altered early Devonian and older rocks pre-dating the main Devonian–Carboniferous sedimentation of south-west England. Such a basement may have included metamorphic rocks overlain by Ordovician quartzites and Silurian limestones. Its rise in the south along a N.E.–S.W. line through the Lizard could have both provided the exotic material of the Meneage crush zone and imparted a Caledonoid trend to south-west Cornwall. Movements at the end of Devonian times, suggested by radiometric dating, may point to the thrusting north-eastwards of a Lizard nappe.

We have noted that east–west normal faults developed as the compressive forces of the Armorican orogeny waned. It is fractures of this type which bound the Permian outlier at Hollacombe, near Holsworthy, and much of the major trough of Permian sediments which runs eastwards from Exbourne through Crediton. Boreholes at Hollacombe proved up to 391 ft of Permian sediments, and gravity surveys of the Crediton trough have indicated up to 1500 ft of similar sediments overlying Carboniferous rocks.

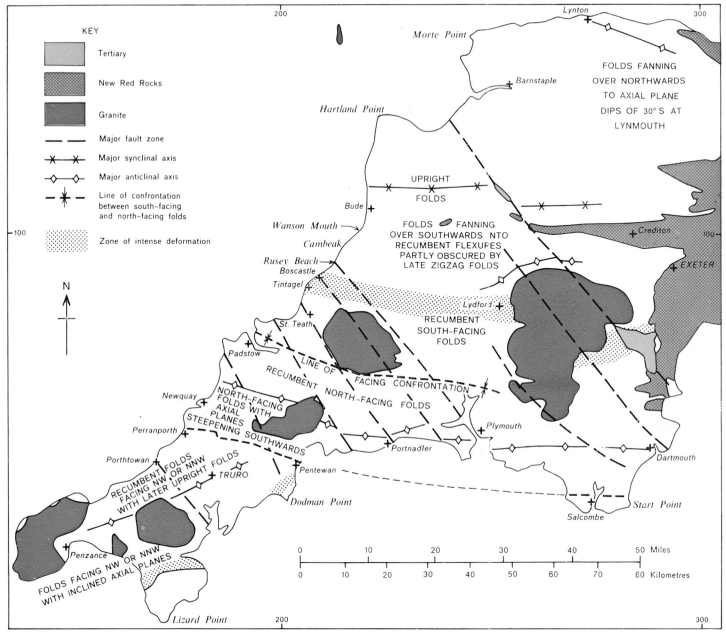

N

Lynton

Morte Point

Barnstaple

Hartland Point

FOLDS FANNING OVER NORTHWARDS TO AXIAL PLANE DIPS OF 30° S AT LYNMOUTH

UPRIGHT FOLDS

Bude

Wanson Mouth

Cambeak

FOLDS FANNING OVER SOUTHWARDS INTO RECUMBENT FLEXURES PARTLY OBSCURED BY LATE ZIGZAG FOLDS

Crediton

EXETER

Rusey Beach
Boscastle

Tintagel

Lydford

RECUMBENT SOUTH-FACING FOLDS

St. Teath

Padstow

LINE OF FACING CONFRONTATION

RECUMBENT NORTH-FACING FOLDS

Newquay

NORTH-FACING FOLDS WITH AXIAL PLANES STEEPENING SOUTHWARDS

Perranporth

Porthtowan

RECUMBENT FOLDS FACING NW OR NNW WITH LATER UPRIGHT FOLDS

TRURO

Pentewan

Portnadler

Plymouth

Dartmouth

Dodman Point

FOLDS FACING NW OR NNW WITH INCLINED AXIAL PLANES

Penzance

Salcombe

Start Point

| 0 | 10 | 20 | 30 | 40 | 50 Miles |

| 0 | 10 | 20 | 30 | 40 | 50 | 60 | 70 | 80 Kilometres |

Lizard Point

FIG. 31. *Structure of south-west England*

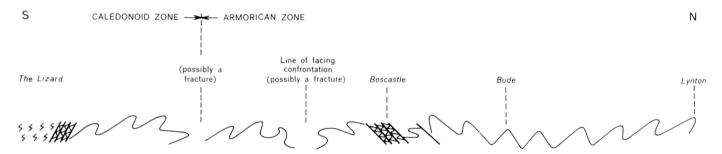

S   CALEDONOID ZONE →← ARMORICAN ZONE   N

The Lizard   (possibly a fracture)   Line of facing confrontation (possibly a fracture)   *Boscastle*   *Bude*   *Lynton*

FIG. 32. *Fold attitudes displayed on the coast*

The great wrench faults which course north-westwards across south-west England have been shown (Dearman 1964b) to have a cumulative dextral displacement of at least 21 miles. Removing this effect places the southern margins of the Dartmoor and Bodmin Moor granites on the same latitude and suggests a slightly more regular original shape for the underlying batholith.

Reid, contributing to the Newton Abbot memoir (1913), considered that the Tertiary sediments of Bovey Tracey were preserved in a fault trough trending north-westwards. Fasham (1971) has estimated the maximum thickness of post-Palaeozoic sediments in the Bovey basin as about 4000 ft. This wrench fault zone, the Sticklepath Fault, displaces the Permian outcrops near Hatherleigh, incorporates the Petrockstow basin (p. 78) and emerges on the north Devon coast in Cockington Cliff. Other similar fault zones are reflected in the shape of the Dartmoor Granite, and more have been located farther west running from Cambeak to Plymouth, Boscastle to Portwrinkle, St. Teath to Portnadler, St. Minver to Lantivet, St. Eval to St. Gorran and Porthtowan to Falmouth Bay.

The relatively simple structures of the Permian, Jurassic and Cretaceous strata are mentioned as part of the stratigraphical accounts in Chapters 6, 7 and 8. In addition, Durrance and Hamblin (1969) and Hamblin (1972) have identified intra-Cretaceous and younger folding and faulting in the Haldon Hills.

Rock	Location	Mineral	Method	Age (million years)	Remarks
Granite	Lundy Island	Biotite	K–Ar	50±3 to 54±2	Granite early Tertiary, probably middle Eocene
		Feldspar and quartz	K–Ar	55±3	
Upper Greensand	Haldon Hills	Glauconite	K–Ar	91	Basal Cenomanian
	Lyme Regis	Glauconite	K–Ar	96	Upper Albian
Phonolite	Wolf Rock	Whole rock	K–Ar	111±6 to 115±7	} Possible argon leakage
	Wolf Rock	Sanidine	K–Ar	113±7	
	Wolf Rock	Sanidine	Ar–Ar	129±3 to 134±2	
	Wolf Rock	Nepheline	K–Ar	262	Lower Permian
Phonolitic lava	Epson Shoal	Whole rock	K–Ar	127±7 to 132±8	Lower Cretaceous
Feldspar vein	The Lizard	Adularia	Rb–Sr	167±8	Mid-Jurassic
Mineral lodes	Wheal Owles and South Terras	Pitchblende, coffinite	U–Pb	47±2 to 71±12	Tertiary mineralization
	Redruth area	Pitchblende	U–Pb	124±4 to 130±7	Apparently Jurasso-Cretaceous activity
	Wheal Bray	Pitchblende	U–Pb	165±4	Mid-Jurassic
	South Terras and Geevor	Pitchblende, uraninite	U–Pb	223±5 to 229±12	Permo-Triassic mineralization
	Geevor	Tourmaline	Ar–Ar	229	
	Various mines, Devon and Cornwall	Galena	Pb–Pb	280±20	} Armorican mineralization
	Geevor and South Crofty	Uraninite	U–Pb	277±11 to 308±30	
	Knap Down and Combe Martin	Galena	Pb–Pb	360±30	Devonian mineralization
Basalt ⎱ 'Exeter	Dunchideock	Biotite	K–Ar	281±11	
Minette ⎰ Traps'	Killerton Park	Biotite	K–Ar	274±10 to 287±11	
Aplite	Meldon	Muscovite	K–Ar	254±5 to 257	
Pegmatite	Giant's Head, Scilly Isles	Muscovite	K–Ar	298±5	
Pegmatoid granite	Godolphin	Muscovite	K–Ar	280±10	
Granite (greisen)	St. Agnes	Muscovite	K–Ar	278±5	
Granite	Dartmoor	Biotite	K–Ar	254 to 295	Armorican
	Dartmoor	Biotite	Rb–Sr	267±8 to 292±12	
	Kit Hill	Biotite	K–Ar	296±8	
	Kit Hill	Muscovite	K–Ar	265±6	
	Bodmin Moor	Biotite	K–Ar	272 to 278±4	
	Bodmin Moor	Muscovite	K–Ar	268±9	
	Haig Fras	Biotite	K–Ar	275	
	Haig Fras	Muscovite	K–Ar	277±10	
	St. Austell	Biotite	K–Ar	277±7	
	St. Austell	Whole rock	Rb–Sr	288±13	
	Carnmenellis	Biotite	K–Ar	275 to 281±9	
	Carnmenellis	Muscovite	K–Ar	280±6	
	Godolphin	Muscovite	K–Ar	289±7	
	Land's End	Biotite	K–Ar	250±15 to 330	
	Land's End	Biotite	Rb–Sr	270±9	
	Scilly Isles	Biotite	K–Ar	303±9	
	Scilly Isles	Muscovite	K–Ar	283±7 to 298±5	
	Seven Stones	Muscovite	K–Ar	281±9	
Dolerite	Meldon	Biotite	K–Ar	295±6 & 296±12	Age of metamorphism
	Lower Ashton	Biotite	K–Ar	337±27	
	Newlyn	Biotite	K–Ar	372±22	
Slate and phyllite	Gunwalloe Cove	Whole rock	K–Ar	345 to 355	⎱ Upper Devonian– Lower Carboniferous Mylor Slate; probably much argon lost
	Hemmick Beach	Whole rock	K–Ar	349 to 356	⎰
	Porthleven	Whole rock	K–Ar	< 300	
Schist	The Lizard	Muscovite	K–Ar	348 to 359	Kennack Gneiss, probably 360 to 390 million years old and much argon lost from metamorphic rocks about this (Caledonian) time. The oldest rocks, hornblende-schist and granulite, may have undergone Pre-Cambrian metamorphism
Schist and gneiss	The Lizard	Biotite	K–Ar	360 to 397	
Schist and gneiss	The Lizard	Biotite	Rb–Sr	352 to 353	
Schist gneiss, amphibolite and granulite	The Lizard	Hornblende	K–Ar	357±20 to 492±26	
Gneiss	Eddystone Rock	Biotite	K–Ar	375±17	

APPENDIX. *Isotopic age-determinations of rocks of south-west England*

List of Geological Survey Maps and Memoirs, and Short Bibliography of Other Works

Maps

(a) **Quarter-inch to one mile**

Sheet 18 with part of 17. Swansea, Cardiff, Bristol, Taunton, Barnstaple. (Out of print.)

,, 21 and 25. Bodmin, Truro, Falmouth, Land's End, Isles of Scilly.

,, 22. Plymouth, Exeter, Lyme Regis.

(b) **One inch to one mile**

  (i) *Old Series Sheets, hand-coloured solid edition (Out of print)*

Sheet 20. Bridgwater, Porlock, etc. (1834, Revised 1839.)[1]

,, 21. Taunton, Wiveliscombe, Dulverton, Bampton, Tiverton, Crediton,Honiton and Axminster. (1834, Revised 1839.)[1]

,, 22. Coast from Lyme Regis to Torbay, Exeter, Newton Abbot, Totnes. (1834, Revised 1839.)[1]

,, 23. Coast from Berry Head to Start Point, Brixham, Dartmouth, etc. (1834, Revised 1839.)[1]

,, 24. Coast from Bolt Head to Looe Bay, Plymouth, Kingsbridge, etc. (1835, Revised 1839.)[1]

,, 25. Launceston, Tavistock, Moretonhampstead, Ashburton, Dartmoor, Callington, and the lodes of the mining district. (1835, Revised 1839, Additions 1866.)[1]

,, 26. Bideford Bay, Torrington, South Molton, Holsworthy, and Hatherleigh. (1835, Revised 1839.)[1]

,, 27. Coast from Foreland Point to Barnstaple, Ilfracombe, Woolacombe, etc. (1835, Revised 1839.)

,, 28. Lundy Island. (1839.)

,, 29. Coast from Hartland Point to Cambeak, Bude Bay. (1839.)[1]

,, 30. Coast from Tintagel to Newquay; Padstow, Camelford, St. Columb, the mining districts of Bodmin, Lostwithiel, etc. (1839, Additions 1866.)[1]

,, 31. South Coast, from Polperro to St. Keverne: North Coast, St. Agnes Head to Portreath, Truro, Falmouth, the mining districts of St. Austell, Fowey, Camborne, Redruth, etc. showing lodes, etc. (1839, Additions 1866.)[1]

,, 32. Lizard Head (the Serpentine District). (1839.)[1]

,, 33. Land's End, Mount's Bay, Penzance, St. Ives Bay, the mining districts of St. Just, Marazion, etc. (1839, Additions 1866.)[1]

---

[1]Partly or wholly replaced by New Series Sheets.

(ii) *New Series Sheets, colour-printed drift edition*

Sheet 294.   Dulverton. (Provisional edition.)
  „   295.   Taunton and Bridgwater.
  „   310.   Tiverton. (Provisional edition.)
  „   311.   Wellington and Chard.
  „   322.   Boscastle.
  „   323.   Holsworthy.
  „   324.   Okehampton.
  „   325.   Exeter.
  „   326 and 340.   Sidmouth and Lyme Regis.
  „   335.   Trevose Head.
  „   336.   Camelford.
  „   337.   Tavistock.
  „   338.   Dartmoor Forest.
  „   339.   Teignmouth.
  „   346.   Newquay.
  „   347.   Bodmin and St. Austell.
  „   348.   Plymouth and Liskeard.
  „   349.   Ivybridge.
  „   350.   Torquay.
  „   351 and 358.   Penzance.
  „   352.   Falmouth, Truro, Camborne and Redruth.
  „   353 and 354.   Mevagissey.
  „   355.   Kingsbridge.
  „   356.   Start Point.
  „   357 and 360.   Isles of Scilly.
  „   359.   Lizard.

(c) Six inches to one mile
  Six-inch maps, where available, may be consulted in the library of the Institute of Geological Sciences. Uncoloured photocopies may be supplied on special order.

(d) Aeromagnetic Map (scale 1 : 250 000)
      Sheet 1.   South-western Approaches.

Memoirs

(a) General Memoirs
  Report on the Geology of Cornwall, Devon and West Somerset, by Sir H. T. De la Beche. (1839.)
  Figures and Descriptions of the Palaeozoic Fossils of Cornwall and West Somerset, by J. Phillips. (1841.)
  Pliocene Deposits of Britain, by C. Reid. (1890.)
  The Jurassic Rocks of Britain, vol. iii, Lias of England and Wales, by H. B. Woodward. (1893.)
  The Cretaceous Rocks of Britain, by A. J. Jukes-Browne, 3 vols. (1900–4.)
  *Chemical Analyses of Igneous Rocks, Metamorphic Rocks and Minerals, by E. M. Guppy (1931) and (for 1931–54) by E. M. Guppy and P. A. Sabine. (1965.)

(b) New Series Sheet Memoirs
      295.   Geology of the Quantock Hills and of Taunton and Bridgwater, by W. A. E. Ussher. (1908.)
      311.   Geology of the Country between Wellington and Chard, by W. A. E. Ussher, with contributions by H. B. Woodward and A. J. Jukes-Browne. (1906.)

(b) **New Series Sheet Memoirs**—*continued*

*322. Geology of the Coast between Tintagel and Bude, by E. C. Freshney, M. C. McKeown and M. Williams. (1972.)

*322 and 323. Geology of the Country around Boscastle and Holsworthy, by M. C. McKeown, E. A. Edmonds, M. Williams, E. C. Freshney and D. J. Masson Smith. (1973.)

*324. Geology of the Country around Okehampton, by E. A. Edmonds, J. E. Wright, K. E. Beer, J. R. Hawkes, M. Williams, E. C. Freshney and P. J. Fenning, with contributions by W. H. C. Ramsbottom, M. C. McKeown and P. E. R. Lovelock. (1968.)

325. Geology of the Country around Exeter, by W. A. E. Ussher, with notes by J. J. H. Teall. (1902.)

326 and 340. Geology of the Country near Sidmouth and Lyme Regis, by H. B. Woodward and W. A. E. Ussher, with contributions by A. J. Jukes-Browne. (1906; 2nd Edition 1911.)

335 and 336. Geology of the Country around Padstow and Camelford, by C. Reid, G. Barrow and H. Dewey, with contributions by J. S. Flett and D. A. MacAlister. (1910.)

337. Geology of the Country around Tavistock and Launceston, by C. Reid, G. Barrow, R. L. Sherlock, D. A. MacAlister and H. Dewey. (1911.)

338. Geology of Dartmoor, by C. Reid, G. Barrow, R. L. Sherlock, D. A. MacAlister, H. Dewey and C. N. Bromehead, with notes by J. S. Flett and W. A. E. Ussher. (1912.)

339. Geology of the Country around Newton Abbot, by W. A. E. Ussher, with contributions by C. Reid, J. S. Flett and D. A. MacAlister. (1913.)

346. Geology of the Country near Newquay, by C. Reid and J. B. Scrivenor, with contributions by J. S. Flett, W. Pollard and D. A. MacAlister. (1906.)

347. Geology of the Country around Bodmin and St. Austell, by W. A. E. Ussher, G. Barrow and D. A. MacAlister, with notes by J. S. Flett. (1909.)

348. Geology of the Country around Plymouth and Liskeard, by W. A. E. Ussher, with notes by J. S. Flett. (1907.)

349. Geology of the Country around Ivybridge and Modbury, by W. A. E. Ussher, with a chapter by G. Barrow. (1912.)

350. Geology of the Country around Torquay, by W. A. E. Ussher; 2nd Edition by W. Lloyd, with notes by C. P. Chatwin and W. G. Shannon. (1903; 2nd Edition 1933.)

351 and 358. Geology of the Land's End District, by C. Reid and J. S. Flett, with notes by B. S. N. Wilkinson, E. E. L. Dixon and W. Pollard and Mining Appendix by D. A. MacAlister. (1907.)

352. Geology of Falmouth and Truro, and of the Mining District of Camborne and Redruth, by J. B. Hill and D. A. MacAlister, with notes by J. S. Flett. (1906.)

353 and 354. Geology of the Country around Mevagissey, by C. Reid, with notes by J. J. H. Teall. (1907.)

355 and 356. Geology of the Country around Kingsbridge and Salcombe, by W. A. E. Ussher. (1904.)

357 and 360. Geology of the Isles of Scilly, by G. Barrow, with contributions by J. S. Flett. (1906.)

*359. Geology of the Lizard and Meneage, by J. S. Flett and J. B. Hill; 2nd Edition by Sir J. S. Flett. (1912; 2nd Edition 1946.)

(c) **Economic Geology** (including **Special Reports on Mineral Resources**)

Vol. I.   Tungsten and Manganese Ores, by Henry Dewey and H. G. Dines, with contributions by C. N. Bromehead, T. Eastwood, G. V. Wilson and R. W. Pocock. (1915; 3rd Edition 1923.)

Vol. II.   Barytes and Witherite, by G. V. Wilson, T. Eastwood, R.W. Pocock and others (1915; 3rd Edition 1922.)

Vol. IV.   Fluorspar, by R. G. Carruthers and R. W. Pocock, with contributions by D. A. Wray, H. Dewey and C. E. N. Bromehead; 4th Edition by K. C. Dunham. (1916; 4th Edition 1950.)

Vol. VII.   Lignites, Jets, Cannel Coals, by Sir A. Strahan and others. (1918; 2nd Edition 1920.)

Vol. IX.   Iron Ores: Somerset, Devon and Cornwall, by T. C. Cantrill, R. L. Sherlock and H. Dewey. (1919.)

Vol. XIV.   Refractory Materials: Fireclays. Resources and Geology, by T. C. Cantrill and others. (1920.)

Vol. XV.   Arsenic and Antimony Ores, by H. Dewey, with contributions by J. S. Flett and G. V. Wilson. (1920.)

Vol. XXI.   Lead, Silver-Lead and Zinc Ores of Cornwall, Devon and Somerset, by Henry Dewey. (1921.)

Vol. XXVII.   Copper Ores of Cornwall and Devon, by Henry Dewey. (1923.)

Vol. XXVIII.   Refractory Materials: Fireclays. Analyses and Physical Tests, by F. R. Ennos and Alexander Scott. (1924.)

Vol. XXXI.   Ball Clays, by Alexander Scott. (1929.)

Kaolin, China-clay and China-stone, Handbook to collection of, in the Museum of Practical Geology, by J. Allen Howe, with an Appendix by Allan B. Dick. (1914.)

Minerals of the British Isles, by F. W. Rudler. (1905.)

*The Metalliferous Mining Region of South-West England, by H. G. Dines, with notes by J. Phemister. 2 vols. (1956.)

*Sources of Road Aggregate in Great Britain (Joint publication of the Road Research Laboratory and the Geological Survey and Museum). 4th Edition (1968).

*Geology and Ceramics, by P. J. Adams. (1961.)

(d) **Bulletins of the Geological Survey**

No. 2—V.   Some Devonian and supposed Ordovician Fossils from South-West Cornwall, by C. J. Stubblefield, pp. 63–71. (1939.)

*No. 6—V.   A Gravimeter Survey of the Ston Easton–Harptree District, East Somerset, by W. Bullerwell, pp. 36–56. (1954.)

*No. 7—IV.   North Exmoor Floods, August 1952, by G. W. Green, pp. 68–84. (1955.)

*No. 29—III.   The Permian Igneous Rocks of Devon, by Diane C. Knill, pp. 115–38. (1969.)

*No. 30—III.   The Cretaceous Structure of Great Haldon, Devon, by E. M. Durrance and R. J. O. Hamblin, pp. 71–88. (1969.)

*No. 41—IV.   The Watchet Fault—a post-Liassic transcurrent reverse fault, by A. Whittaker, pp. 75–80. (1972.)

*No. 42—IV.   Drainage chronology of the South Molton area, Devonshire, by E. A. Edmonds, pp. 99–104. (1972.)

All the memoirs listed above are out of print, except those marked by an asterisk.

Short Bibliography of Other Works

ALLAN, T. D. 1961. A magnetic survey in the western English channel. *Quart. J. Geol. Soc.*, **117**, 157–70.

ANNIS, L. G. 1933. The Upper Devonian rocks of the Chudleigh area, South Devon. *Quart. J. Geol. Soc.*, **89**, 431–47.

ARBER, E. A. N. 1907. On the Upper Carboniferous rocks of west Devon and north Cornwall. *Quart. J. Geol. Soc.*, **63**, 1–28.

ARBER, MURIEL A. 1940. The coastal landslips of south-east Devon. *Proc. Geol. Assoc.*, **51**, 257–71.

—— 1949. Cliff profiles of Devon and Cornwall. *Geog. J.*, **114**, 191–7.

ASHWIN, D. P. 1958. The coastal outcrop of the Culm Measures of south-west England. *Abs. Proc. Conf. Geol. and Geomorph. in S.W. England, Roy. Geol. Soc. Corn.*, 2.

BARTON, R. M. 1964. *Introduction to the Geology of Cornwall*. Truro.

BATSTONE, A. E. 1959. The structure and tectonic history of the Tintagel–Davidstowe area. *Trans. Roy. Geol. Soc. Corn.*, **19**, 17–32.

BLUNDELL, D. J. 1957. A palaeomagnetic investigation of the Lundy dyke swarm. *Geol. Mag.*, **94**, 187–93.

BLYTH, F. G. H. 1957. The Lustleigh Fault in north-east Dartmoor. *Geol. Mag.*, **94**, 291–6.

—— 1962. The structure of the north-eastern tract of the Dartmoor Granite. *Quart. J. Geol. Soc.*, **118**, 435–53.

BONNEY, T. G. 1877. On the microscopic structure of luxullianite. *Miner. Mag.*, **1**, 215–21.

BORLASE, W. 1758. *The Natural History of Cornwall*. Oxford.

BOTT, M. H. P., DAY, A. A. and MASSON SMITH, D. 1958. The geological interpretation of gravity and magnetic surveys in Devon and Cornwall. *Phil. Trans. Roy. Soc.* (A), **251**, 161–91.

BRADSHAW, J. D., RENOUF, J. T. and TAYLOR, R. T. 1967. The development of Brioverian structures and Brioverian/Palaeozoic relationships in west Finistère (France). *Geol. Rdsch.*, **56**, 567–96.

BRAITHWAITE, C. J. R. 1966. The petrology of Middle Devonian limestones in south Devon, England. *J. sedim. Petrol.*, **36**, 176–92.

BRAMMALL, A. and HARWOOD, H. F. 1932. The Dartmoor Granites. *Quart. J. Geol. Soc.*, **88**, 171–237.

BROUGHTON, D. G. 1967. Tin working in the eastern district of the parish of Chagford, Devon. *Proc. Geol. Assoc.*, **78**, 447–62.

BUCKLAND, W. 1822. On the excavation of valleys by diluvian action, as illustrated by a succession of valleys which intersect the south coast of Dorset and Devon. *Trans. Geol. Soc.* (2), **1**, 95–102.

BUTCHER, N. E. and HODSON, F. 1960. A review of the Carboniferous goniatite zones in Devon and Cornwall. *Palaeontology*, **3**, pt. 1, 75–81.

CHAMPERNOWNE, A. and USSHER, W. A. E. 1879. Notes on the structure of the Palaeozoic districts of west Somerset. *Quart. J. Geol. Soc.*, **25**, 532–48.

COLLINS, J. H. 1912. Observations on the west of England mining region. *Trans. Roy. Geol. Soc. Corn.*, **14**.

CONYBEARE, J. J. 1814. Memoranda relative to Clovelly, north Devon. *Trans. Geol. Soc.*, **2**, ser. 1, 498.

—— 1823. On the geology of Devon and Cornwall. *Ann. Phil.*, ser. 2, **5**, 184; **6**, 35.

CORNWELL, J. D. 1967. Palaeomagnetism of the Exeter lavas, Devonshire. *Geophys. J.*, **12**, 181–96.

CROOKALL, R. 1930. The plant horizons represented in the Barren Coal Measures of Devon, Cornwall and Somerset. *Proc. Cotteswold Nat. Field Club*, **24**, 27–34.

CURRY, D., MARTINI, E., SMITH, A. J. and WHITTARD, W. F. 1962. The geology of the western approaches of the English Channel. I. Chalky rocks from the upper reaches of the continental slope. *Phil. Trans. Roy. Soc.* (B), **245**, 267–90.

—— HERSEY, J. B., MARTINI, E. and WHITTARD, W. F. 1965. The geology of the western approaches of the English Channel. II. Geological interpretation aided by boomer and sparker records. *Phil. Trans. Roy. Soc.* (B), **248**, 315–51.

DAVISON, E. H. 1926. A study of the Cornish granite, its variation and its relation with the occurrence of tin and other metallic ores. *Trans. Roy. Geol. Soc. Corn.*, **15**, 578–92.

—— 1927. Recent evidence confirming the zonal arrangement of minerals in the Cornish lodes. *Econ. Geol.*, **22**, 475–9.

DE RAAF, J. F. M., READING, H. G. and WALKER, R. G. 1965. Cyclic sedimentation in the lower Westphalian of north Devon, England. *Sedimentology*, **4**, 1–52.

DEARMAN, W. R. 1959. The structure of the Culm Measures at Meldon, near Okehampton, north Devon. *Quart. J. Geol. Soc.*, **115**, 65–106.

—— 1962. Dartmoor. *Geol. Assoc.* Guide No. 33.

—— 1964a. Dartmoor: its geological setting. *Dartmoor Essays, Devon. Assoc.*

—— 1964b. Wrench-faulting in Cornwall and south Devon. *Proc. Geol. Assoc.*, **74**, 265–87.

—— and BUTCHER, N. E. 1959. The geology of the Devonian and Carboniferous rocks of the north-west border of the Dartmoor Granite, Devonshire. *Proc. Geol. Assoc.*, **70**, 51–92.

—— and FRESHNEY, E. C. 1967. Repeated folding at Boscastle, north Cornwall, England. *Proc. Geol. Assoc.*, **77**, 199–215.

DEWEY, H. 1925. The mineral zones of Cornwall. *Proc. Geol. Assoc.*, **36**, 107–35.

—— 1935. South-west England. *Brit. Reg. Geol. Geol.Surv.*

DINELEY, D. L. 1961. The Devonian System in south Devonshire. *Fld. Stud.*, **1**, 121–40.

—— 1966. The Dartmouth Beds of Bigbury Bay, south Devon. *Quart. J. Geol. Soc.*, **122**, 187–217.

—— and RHODES, F. H. T. 1956. Conodont horizons in the west and south-west of England. *Geol. Mag.*, **93**, 242–8.

DINES, H. G. 1934. Lateral extent of ore shoots. *Trans. Roy. Geol. Soc. Corn.*, **16**, 279–96.

DODSON, M. H. 1961. Isotopic ages from the Lizard peninsula, south Cornwall. *Proc. Geol. Soc.*, No. 1591, 133–6.

DOLLAR, A. T. J. 1942. The Lundy complex: its petrology and tectonics. *Quart. J. Geol. Soc.*, **97**, 39–77.

ELLIOT, G. F. 1961. A new British Devonian alga, *Palaeoporella lummatonensis*, and the brachiopod evidence of the age of the Lummaton Shell-bed. *Proc. Geol. Assoc.*, **72**, 251–60.

EL SHARKAWI, M. A. H. and DEARMAN, W. R. 1966. Tin-bearing skarns from the north-west border of the Dartmoor Granite, Devonshire, England. *Econ. Geol.*, **61**, 362–9.

ERBEN, H. K. 1964. Facies developments in the marine Devonian of the Old World. *Proc. Ussher Soc.*, **1**, 92–118.

EVANS, J. W. 1922. The geological structure of the country around Combe Martin, north Devon. *Proc. Geol. Assoc.*, **33**, 201–28.

EXLEY, C. S. 1959. Magmatic differentiation and alteration in the St. Austell Granite. *Quart. J. Geol. Soc.*, **114**, 197-230.

—— and STONE, M. [1966]. The granitic rocks of south-west England. *In* Present views of some aspects of the geology of Cornwall and Devon. *Roy. Geol. Soc. Corn.*, Commem. Vol. for 1964, 131–84.

FITCH, F. J. and MILLER, J. A. 1964. The age of the paroxysmal Variscan orogeny in England. *In* The Phanerozoic Time-scale. *Quart. J. Geol. Soc.*, **120S**, Symposium Volume, 159–75.

FLOYD, P. A. 1965. Metasomatic hornfelses of the Land's End aureole at Tater-du, Cornwall. *J. Petrology*, **6**, 223–45.

FOX, H. 1895. Several papers in *Trans. Roy. Geol. Soc. Corn.*, **11**.

—— 1896. The radiolarian cherts of Cornwall. *Trans. Roy. Geol. Soc. Corn.*, **12**, 39.

FYSON, W. K. 1962. Tectonic structures in the Devonian rocks near Plymouth, Devon. *Geol. Mag.*, **99**, 208–26.

GARNETT, R. H. T. 1961. Structural control of mineralization in south-west England. *Mining Mag.*, **105**, 329–37.

GARROD, D. A. E. 1926. *The Upper Palaeolithic Age in Britain*. Oxford.

GODWIN-AUSTEN, R. A. C. 1842. On the geology of the south-east of Devonshire. *Trans. Geol. Soc.*, (2) **6**, 433–9.

GOLDRING, R. 1955. The Upper Devonian and Lower Carboniferous trilobites of the Pilton Beds in north Devon. *Senckenbergiana lethea*, **36**, 27–48.

—— 1962. The bathyal lull: Upper Devonian and Lower Carboniferous sedimentation in the Variscan geosyncline. *Some Aspects of the Variscan Fold Belt*, 75–91. Manchester.

—— 1971. Shallow-water Sedimentation (as illustrated in the Upper Devonian Baggy Beds). *Mem. Geol. Soc. Lond.*, **5**.

GREEN, D. H. 1964a. The petrogenesis of the high temperature peridotite in the Lizard area, Cornwall. *J. Petrology*, **5**, 134–88.

—— 1964b. The metamorphic aureole of the peridotite at the Lizard, Cornwall. *J. Geol.*, **72**, 543–63.

—— [1966]. A re-study and re-interpretation of the geology of the Lizard peninsula, Cornwall. *In* Present views of some aspects of the geology of Cornwall and Devon. *Roy. Geol. Soc. Corn.*, Commem. Vol. for 1964, 87–114.

GROVES, A. W. 1952. Wartime investigations into the haematite and manganese ore resources of Great Britain and Northern Ireland. *Min. Supply Monographs*, No. 20, 703.

HALLAM, A. 1960. The White Lias of the Devon coast. *Proc. Geol. Assoc.*, **71**, 47–60.

HENDRIKS, E. M. LIND. 1931. The stratigraphy of south Cornwall. *Rep. Brit. Assoc.*, (Bristol 1930), 332.

—— 1937. Rock succession in south Cornwall, *Quart. J. Geol. Soc.*, **93**, 322–67.

—— 1939. The Start-Dodman-Lizard boundary zone in relation to the Alpine structure of Cornwall. *Geol. Mag.*, **76**, 385–401.

—— 1949. The Gramscatho Series. *Trans. Roy. Geol. Soc. Corn.*, **18**, 50–64.

—— 1951. Geological succession and structure in western south Devonshire. *Trans. Roy. Geol. Soc. Corn.*, **18**, 255–308.

—— 1959. A summary of the present views on the structure of Devon and Cornwall. *Geol. Mag.*, **96**, 253–7.

HENWOOD, W. J. 1843. On the metalliferous deposits of Cornwall and Devon. *Trans. Roy. Geol. Soc. Corn.*, **5**.

HICKS, H. 1896. The Morte Slates. *Quart. J. Geol. Soc*, **52**, 254–72.

HILL, M. N. and KING, W. B. R. 1953. Seismic prospecting in the English Channel and its geological interpretation. *Quart. J. Geol. Soc.*, **109**, 1–19,

HILL, M. N. and VINE, F. J. 1965. A preliminary magnetic survey of the western approaches of the English Channel. *Quart. J. Geol. Soc.*, **121**, 463–76.

HOLL, H. B. 1868. On the older rocks of south Devon and east Cornwall. *Quart. J. Geol. Soc.*, **24**, 400–34.

HOLWILL, F. J. W. 1962. The succession of limestones within the Ilfracombe Beds (Devonian) of north Devon. *Proc. Geol. Assoc.*, **73**, 281–93.

HOSKING, K. F. G. 1952. Cornish pegmatites and bodies with pegmatite affinity. *Trans. Roy. Geol. Soc. Corn.*, **18**, 411–55.

—— [1966]. Permo-Carboniferous and later primary mineralization of Cornwall and south-west Devon. *In* Present views of some aspects of the geology of Cornwall and Devon. *Roy. Geol. Soc. Corn.*, Commem. Vol. for 1964, 201–45.

—— and TROUNSON, J. H. 1959. The mineral potential of Cornwall. *In* The future of non-ferrous mining in Great Britain and Ireland. *Instn Min. Metall. Lond.*, 335–69.

HOUSE, M. R. 1956. Devonian goniatites from north Cornwall. *Geol. Mag.*, **93**, 257–62.

—— 1960. Upper Devonian ammonoids from north-west Dartmoor, Devonshire. *Proc. Geol. Assoc.*, **70**, 315–21.

—— 1963. Devonian ammonoid successions and facies in Devon and Cornwall. *Quart. J. Geol. Soc.*, **119**, 1–27.

—— and BUTCHER, N. E. 1962. Excavations in the Devonian and Carboniferous rocks of the Chudleigh area, south Devon. *Proc. Ussher Soc.*, **1**, pt. 1, 28.

—— and SELWOOD, E. B. [1966]. Palaeozoic palaeontology in Devon and Cornwall. *In* Present views of some aspects of the geology of Cornwall and Devon. *Roy. Geol. Soc. Corn.*, Commem. Vol. for 1964, 45–86.

HUTCHINS, P. F. 1963. The Lower New Red Sandstone of the Crediton valley. *Geol. Mag.*, **100**, 107–28.

IRVING, A. 1888. The red-rock series of the Devon coast-section. *Quart. J. Geol. Soc.*, **44**, 149–63.

KENNARD, A. S. 1945. The early digs in Kent's Hole, Torquay. *Proc. Geol. Assoc.*, **56**, 156–213.

KING, W. B. R. 1954. The geological history of the English Channel. *Quart. J. Geol. Soc.*, **110**, 77–101.

LAMBERT, J. L. M. 1959. Cross-folding in the Gramscatho Beds at Helford River, Cornwall. *Geol. Mag.*, **96**, 489–96.

—— 1965. A re-interpretation of the breccias in the Meneage crush zone of the Lizard boundary, south-west England. *Quart. J. Geol. Soc.*, **121**, 339–58.

LAMING, D. J. C. 1965. Age of the New Red Sandstone in south Devonshire. *Nature*, **207**, 624–5.

—— 1966. Imbrication, palaeocurrents and other sedimentary features in the Lower New Red Sandstone, Devonshire, England. *J. sedim. Petrol*, **36**, pt. 4, 940–59.

LANG, W. D. 1914. The geology of the Charmouth cliffs, beach and foreshore. *Proc. Geol. Assoc.*, **25**, 293–360.

LEESE, C. E. and SETCHELL, J. 1937. Notes on Delabole Slate Quarry. *Trans. Roy. Geol. Soc. Corn.*, **17**, 41–7.

MIDDLETON, G. V. 1960. Spilitic rocks in south-east Devonshire. *Geol. Mag.*, **97**, 192–207.

MILLER, J. A. and GREEN, D. H. 1961. Age determinations of rocks in the Lizard (Cornwall) area. *Nature*, **192**, 1175.

—— and MOHR, P. A. 1964. Potassium-argon measurements on the granites and some associated rocks from south-west England. *Geol. J.*, **4**, 105–26.

MILNER, H. B. 1922. The nature and origin of the Pliocene deposits of the County of Cornwall, and their bearing on the Pliocene geography of the south-west of England. *Quart. J. Geol. Soc.*, **78**, 348–77.

MITCHELL, G. F. 1966. The St. Erth Beds—An alternative explanation. *Proc. Geol. Assoc.*, **76**, 345–66.

MOORE, E. W. J. 1929. The occurrence of *Reticuloceras reticulatum* in the Culm of north Devon, *Geol. Mag.*, **66**, 356–8.

OSMAN, C. W. 1928. The granites of the Scilly Isles and their relation to the Dartmoor granites. *Quart. J. Geol. Soc.*, **84**, 258–92.

OWEN, D. E. 1934. The Carboniferous rocks of the north Cornish coast and their structures. *Proc. Geol. Assoc.*, **45**, 451–71.

—— 1951. Carboniferous deposits in Cornubia. *Trans. Roy. Geol. Soc. Corn.*, **18**, 65–104.

PAUL, H. 1937. The relationship of the Pilton Beds in north Devon to their equivalents on the continent. *Geol. Mag.*, **74**, 433–42.

PEACH, C. W. 1841. An account of the fossil organic remains found on the southeast coast of Cornwall. *Trans. Roy. Geol. Soc. Corn.*, **6**, 12.

—— 1881. On fossils from the rocks of Cornwall. *Trans. Roy. Geol. Soc. Corn.*, **10**, 90.

PENGELLY, W. 1863. On the chronological value of the New Red Sandstone System of Devonshire. *Trans. Devon. Assoc.*, **1**, pt. 2, 31–43.

—— 1864. The denudation of rocks in Devonshire. *Trans. Devon. Assoc.*, **1**, pt. 3, 42–59.

PHILLIPS, F. C. 1928. Metamorphism in the Upper Devonian of north Cornwall. *Geol. Mag.*, **65**, 541–56.

—— 1964. Metamorphic rocks of the sea floor between Start Point and Dodman Point, S.W. England. *J. mar. biol. Assoc. U.K.*, **44**, 655–63.

—— [1966]. Metamorphism in south-west England. *In* Present views of some aspects of the geology of Cornwall and Devon. *Roy. Geol. Soc. Corn.*, Commem. Vol. for 1964, 185–200.

PHILLIPS, J. A. 1876. On the so-called 'Greenstones' of western Cornwall. *Quart. J. Geol. Soc.*, **32**, 155–79.

—— 1878. On the so-called 'Greenstones' of central and eastern Cornwall. *Quart. J. Geol. Soc.*, **34**, 471–97.

PRENTICE, J. E. 1960a. Dinantian, Namurian and Westphalian rocks of the district south-west of Barnstaple, north Devon. *Quart. J. Geol. Soc.*, **115**, 261–89.

—— 1960b. The stratigraphy of the Upper Carboniferous rocks of the Bideford region, north Devon. *Quart. J. Geol. Soc.*, **116**, 397–408.

—— 1962. The sedimentation history of the Carboniferous in Devon. *Some aspects of the Variscan fold belt*, 93–108. Manchester.

—— and THOMAS, J. M. 1960. The Carboniferous goniatites of north Devon. *Abs. Proc. Conf. Geol. and Geomorph. in S.W. England, Roy. Geol. Soc. Corn.*, 6–8.

PRYCE, W. 1778. *Mineralogia Cornubiensis*, a treatise on minerals, mines and mining. London.

READ, H. H. 1957. *The granite controversy*. London.

READING, H. G. 1965. Recent finds in the Upper Carboniferous of south-west England and their significance. *Nature*, **208**, 745–8.

RICHARDSON, L. 1906. On the Rhaetic and contiguous deposits of Devon and Dorset. *Proc. Geol. Assoc.*, **19**, 401–9.

—— 1911. The Rhaetic and contiguous deposits of west, mid and part of east Somerset. *Quart. J. Geol. Soc.*, **67**, 1–74.

RICHTER, D. 1965. Stratigraphy, igneous rocks and structural development of the Torquay area. *Trans. Devon. Assoc.*, **97**, 57–70.

RICHTER, D. 1967. Sedimentology and facies of the Meadfoot Beds (Lower Devonian) in south-east Devon. *Geol. Rdsch.*, **56**, 543–61.

ROBSON, J. 1946. Geology of Carn Brea. *Trans. Roy. Geol. Soc. Corn.*, **17**, 208–20.

—— 1947. The structure of Cornwall. *Trans. Roy. Geol. Soc. Corn.*, **17**, 227–46.

—— 1949. Geology of the Land's End peninsula. *Trans. Roy. Geol. Soc. Corn.*, **17**, 427–54.

ROGERS, I. 1909. On a further discovery of fossil fish and mollusca in the Upper Culm Measures of north Devon. *Trans. Devon. Assoc.*, **41**, 309–19.

—— and SIMPSON, B. 1937. The flint gravel deposit of Orleigh Court, Buckland Brewer, north Devon. *Geol. Mag.*, **74**, 309–16.

ROWE, A. W. and SHERBORN, C. D. 1903. The zones of the white chalk of the English coast, III, Devon. *Proc. Geol. Assoc.*, **18**, 1–52.

SABINE, P. A. 1968. Origin and age of solutions causing the wallrock alteration of the Perran Iron Lode, Cornwall. *Trans. Instn Min. Metall. Lond.*, (B), **77**, B1–5.

—— and WATSON, J. V. 1965 (1967; 1968). Isotopic age-determinations of rocks from the British Isles, 1955–64 (1965; 1966). *Quart. J. Geol. Soc.*, **121** (**122; 123**), 477–533 (433–59; 379–93).

SCRIVENOR, J. B. 1948. The New Red Sandstone of south Devonshire. *Geol. Mag.*, **85**, 317–22.

—— 1949. The Lizard–Start problem. *Geol. Mag.*, **86**, 377–86.

SEDGWICK, A. and MURCHISON, R. I. 1840. On the physical structure of Devonshire, and on the subdivisions and geological relations of its older stratified deposits. *Trans. Geol. Soc.*, **5**, 633–705.

SELWOOD, E. B. 1960. Ammonoids and trilobites from the Upper Devonian and lowest Carboniferous of the Launceston area of Cornwall. *Palaeontology*, **3**, pt. 2, 153–85.

—— 1961. The Upper Devonian and Lower Carboniferous stratigraphy of Boscastle and Tintagel, Cornwall. *Geol. Mag.*, **98**, 162–7.

SHERLOCK, R. L. 1947. *The Permo-Triassic formations.* London.

SIMPSON, S. 1951. Some solved and unsolved problems of the stratigraphy of the marine Devonian in Great Britain. *Abs. Senckenb. naturforsch. Ges.*, **485**, 53–66.

—— 1964. The supposed 690 ft. marine platform in Devon. *Proc. Ussher Soc.*, **1**, pt. 3, 89–91.

SMITH, A. J., STRIDE, A. H., WHITTARD, W. F. and SABINE, P. A. 1965. The geology of the western approaches of the English Channel. IV. A recently discovered Variscan granite west-north-west of the Scilly Isles. *Proc. Seventeenth Symposium Colston Research Soc., Colston Papers.*, **17**, 287–301.

SMITH, W. E. 1957. The Cenomanian Limestone of the Beer district, south Devon. *Proc. Geol. Assoc.*, **68**, 115–38.

—— 1961. The Cenomanian deposits of south-east Devonshire. *Proc. Geol. Assoc.*, **72**, 91–134.

—— and DRUMMOND, P. V. O. 1962. Easter field meeting: The Upper Albian and Cenomanian deposits of Wessex. *Proc. Geol. Assoc.*, **73**, 335–52.

SPARGO, T. 1865. *The Mines of Cornwall and Devon: Statistics and Observations.* London.

STEPHENS, N. 1961. A re-examination of some Pleistocene sections in Cornwall and Devon. *Abstr. Proc. Conf. Geol. and Geomorph. S.W. England, Roy. Geol. Soc. Corn.*, 21–3.

STONE, M. 1966. Fold structures in the Mylor Beds, near Porthleven, Cornwall. *Geol. Mag.*, **103**, 440–60.

—— and AUSTIN, W. G. C. 1961. The metasomatic origin of the potash feldspar megacrysts in the granites of south-west England. *J. Geol.*, **69**, 464–72.

TAYLOR, C. W. 1956. Erratics of the Saunton and Fremington areas. *Trans. Devon. Assoc.*, **88**, 52–64.

TAYLOR, P. W. 1950. The Plymouth Limestone. *Trans. Roy. Geol. Soc. Corn.*, **18**, 146–214.

THOMAS, A. N. 1940. The Triassic rocks of north-west Somerset. *Proc. Geol. Assoc.*, **51**, 1–43.

THOMAS, H. H. 1902. The mineralogical constitution of the finer material of the Bunter Pebble Bed in the west of England. *Quart. J. Geol. Soc.*, **58**, 620–32.

THOMAS, J. M. 1962. The Culm Measures in north-east Devon. *Proc. Ussher Soc.*, **1**, pt. 1, 29.

—— 1963. The Culm Measures succession in north-east Devon and north-west Somerset. *Proc. Ussher Soc.*, **1**, pt. 2, 63–4.

TIDMARSH, W. G. 1932. Permian lavas of Devon. *Quart. J. Geol. Soc.*, **138**, 712–73.

TILLEY, C. E. 1923. The petrology of the metamorphosed rocks of the Start area (south Devon). *Quart. J. Geol. Soc.*, **79**, 172–204.

TRESISE, G. R. 1960. Aspects of the lithology of the Wessex Upper Greensand. *Proc. Geol. Assoc.*, **71**, 316–39.

—— 1961. The nature and origin of chert in the Upper Greensand of Wessex. *Proc. Geol. Assoc.*, **72**, 333–56.

USSHER, W. A. E. 1876. On the Triassic rocks of Somerset and Devon. *Quart. J. Geol. Soc.*, **32**, 367–94.

—— 1878. On the chronological value of the Triassic strata of the south-western counties. *Quart. J. Geol. Soc.*, **34**, 459–69.

—— 1887. The Culm of Devonshire. *Geol. Mag.*(3), **24**, 10–17.

—— 1892. The British Culm Measures. *Proc. Somerset Arch. N.H. Soc.*, **38**, 111–219.

—— 1900. The Devonian, Carboniferous and New Red rocks of west Somerset, Devon and Cornwall. *Proc. Somerset Arch. N.H. Soc.*, **46**, 1–64.

—— 1901. The Culm-Measure types of Great Britain. *Trans. Inst. Mining Eng.*, **20**, 360–91.

—— 1906. Devonshire: *Victoria History of the Counties of England*. London.

VAUGHAN, A. 1904. Notes on the Lower Culm of north Devon. *Geol. Mag.*, **41**, 530–2.

WEBBY, B. D. 1965a. The stratigraphy and structure of the Devonian rocks in the Brendon Hills, west Somerset. *Proc. Geol. Assoc.*, **76**, 39–60.

—— 1965b. The stratigraphy and structure of the Devonian rocks in the Quantock Hills, west Somerset. *Proc. Geol. Assoc.*, **76**, 321–44.

—— 1965c. The Middle Devonian marine transgression in north Devon and west Somerset. *Geol. Mag.*, **102**, 478–88.

—— and THOMAS, J. M. 1965. Whitsun field meeting: Devonian of west Somerset and the Carboniferous of north-east Devon. *Proc. Geol. Assoc.*, **76**, 179–94.

WHITAKER, W. 1869. On the succession of beds in the 'New Red' on the south coast of Devon, and on the locality of a new specimen of *Hyperodapedon*. *Quart. J. Geol. Soc.*, **25**, 152–8.

WILSON, G. 1946. The relationship of slaty cleavage and kindred structures to tectonics. *Proc. Geol. Assoc.*, **57**, 263–300

—— 1951. The tectonics of the Tintagel area, north Cornwall. *Quart. J. Geol. Soc.*, **106**, 393–432.

—— 1952. The influence of rock structures on coastline and cliff development around Tintagel, north Cornwall. *Proc. Geol. Assoc.*, **63**, 20–48.

WORTH, R. N. 1890. The igneous constituents of the Triassic breccias and conglomerates of south Devon. *Quart. J. Geol. Soc.*, **46**, 69–81.

Additional References 1973

BUTCHER, N. E. and HOUSE, M. R. 1972. Excavations in the Upper Devonian and Carboniferous rocks near Chudleigh, south Devon. *Trans. Roy. Geol. Soc. Corn.*, **20**, pt. 3, 199–220.

DEARMAN, W. R. 1969. An outline of the structural geology of Cornwall. *Proc. Geol. Soc.*, 1654, 33–9.

—— 1970. Some aspects of the tectonic evolution of south-west England. *Proc. Geol. Assoc.*, **81**, 483–92.

—— 1971. A general view of the structure of Cornubia. *Proc. Ussher Soc.*, **2**, 220–36.

DODSON, M. H. and REX, D. C. 1971. Potassium-argon ages of slates and phyllites from south-west England. *Quart. J. Geol. Soc.*, **126**, 465–99.

DURRANCE, E. M. and HAMBLIN, R. J. O. 1969. The Cretaceous structure of Great Haldon. *Proc. Ussher Soc.*, **2**, 84.

EDMONDS, E. A. 1972. The Pleistocene history of the Barnstaple area. *Rep. No. 72/2, Inst. geol. Sci.*, 12 pp.

—— 1974. Classification of the Carboniferous rocks of south-west England. *Rep. No. 74/13, Inst. geol. Sci.*

FASHAM, M. J. R. 1971. A gravity survey of the Bovey Tracey basin, Devon. *Geol. Mag.*, **108**, 119–29.

FRESHNEY, E. C. and TAYLOR, R. T. 1971. The structure of mid-Devon and north Cornwall. *Proc. Ussher Soc.*, **2**, 241–8.

—— —— 1973. The Upper Carboniferous stratigraphy of north Cornwall and west Devon. *Proc. Ussher Soc.*, **2**, 464–71.

HALL, T. M. 1867. On the relative distribution of fossils throughout the North Devon Series. *Quart. J. Geol. Soc.*, **23**, 371–80.

HAMBLIN, R. J. O. 1972. The Tertiary structure of the Haldon Hills. *Proc. Ussher Soc.*, **2**, 442–6.

HOLWILL, F. J. W., HOUSE, M. R., LANE, R., GAUSS, G. A., HENDRIKS, E. M. L., and DEARMAN, W. R. 1969. Summer (1966) field meeting in Devon and Cornwall. *Proc. Geol. Assoc.*, **80**, 43–62.

LUCAS, M. D. 1960. The Upper Palaeozoic geology of the Dulverton district. *M.Sc. Thesis, University of Bristol.*

RAMSBOTTOM, W. H. C. and CALVER, M. A. 1962. Some marine horizons containing *Gastrioceras* in north-west Europe. *C.R. 4me Cong. Strat. Carb.* Heerlen, 1958, **3**, 571–6.

ROBERTS, J. L. and SANDERSON, D. J. 1971. Polyphase development of slaty cleavage and the confrontation of facing directions in the Devonian rocks of north Cornwall. *Nature, Lond.* (Physical Sciences), **230**, 87–9.

SABINE, P. A. 1968. Kaolinitic wall-rock alteration of the Perran Iron Lode, Cornwall. *Rep. 23rd Session Int. geol. Congr.*, Prague, **14**, *Proc. Symp.* I. Genesis of the Kaolin Deposits, 45–53.

—— and SNELLING, N. J. 1969. The Seven Stones Granite, between Land's End and the Scilly Isles. *Proc. geol. Soc. Lond.*, No. 1654, 47–50.

SANDERSON, D. J. 1971. Superposed folding at the northern margin of the Gram-scatho and Mylor Beds, Perranporth, Cornwall. *Proc. Ussher Soc.*, **2**, 266–9.

—— and DEARMAN, W. R. 1973. Structural zones of the Variscan fold belt in S.W. England, their location and development. *J. Geol. Soc.*, **129**, 527–33.

SIMPSON, S. 1969. Geology. In *Exeter and its region*. University of Exeter.

—— 1970. The structural geology of Cornwall: some comments. *Proc. Geol. Soc.*, 1662, 1–3.

—— 1971. The Variscan structure of north Devon. *Proc. Ussher Soc.*, **2**, 249–52.

Index

Printed in England for Her Majesty's Stationery Office by Hull Printers Ltd., Great Gutter Lane, Willerby, Hull, Yorks.
Dd. 505168 K160

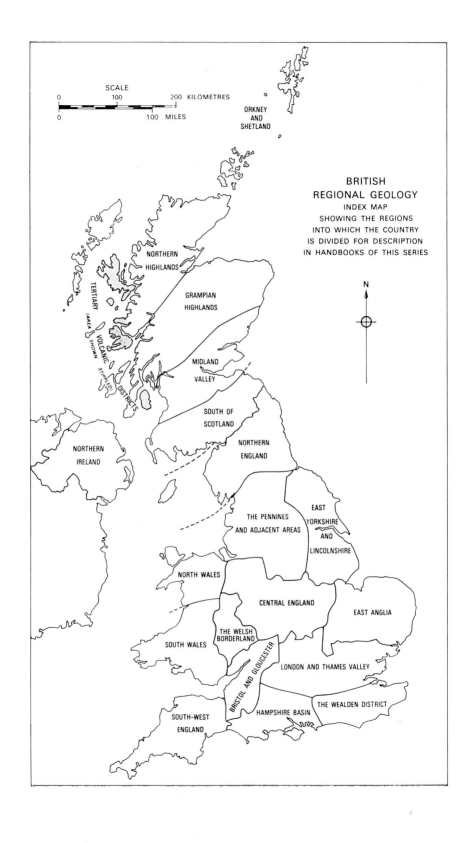

SCALE

0   100   200 KILOMETRES

0   100 MILES

ORKNEY
AND
SHETLAND

BRITISH
REGIONAL GEOLOGY
INDEX MAP
SHOWING THE REGIONS
INTO WHICH THE COUNTRY
IS DIVIDED FOR DESCRIPTION
IN HANDBOOKS OF THIS SERIES

N

NORTHERN
HIGHLANDS

GRAMPIAN
HIGHLANDS

TERTIARY VOLCANIC DISTRICTS
(AREA SHOWN STIPPLED)

MIDLAND
VALLEY

SOUTH OF
SCOTLAND

NORTHERN
IRELAND

NORTHERN
ENGLAND

EAST
YORKSHIRE
AND
LINCOLNSHIRE

THE PENNINES
AND ADJACENT AREAS

NORTH WALES

CENTRAL ENGLAND

EAST ANGLIA

THE WELSH
BORDERLAND

SOUTH WALES

BRISTOL AND GLOUCESTER

LONDON AND THAMES VALLEY

THE WEALDEN DISTRICT

SOUTH-WEST
ENGLAND

HAMPSHIRE BASIN

List of Handbooks on the Regional Geology of Great Britain

England and Wales

Northern England (*4th Edition*, 1971)	40p
London and Thames Valley (*3rd Edition*, 1960)	30p
The Wealden District (*4th Edition*, 1965)	30p
Central England District (*3rd Edition*, 1969)	40p
East Yorkshire and Lincolnshire (1948)	45p
The Welsh Borderland (*3rd Edition*, 1971)	40p
Hampshire Basin and Adjoining Areas (*3rd Edition*, 1960)	30p
East Anglia and Adjoining Areas (*4th Edition*, 1961)	30p
South Wales (*3rd Edition*, 1970)	50p
North Wales (*3rd Edition*, 1961)	30p
The Pennines and Adjacent Areas (*3rd Edition*, 1954)	45p
Bristol and Gloucester District (*2nd Edition*, 1948)	45p

Scotland

Grampian Highlands (*3rd Edition*, 1966)	50p
Northern Highlands (*3rd Edition*, 1960)	35p
South of Scotland (*3rd Edition*, 1971)	50p
Midland Valley of Scotland (*2nd Edition*, 1948)	32½p
Tertiary Volcanic Districts (*3rd Edition*, 1961)	35p
Orkney and Shetland	in press

The above prices do not include postage

The handbooks are obtainable from the Institute of Geological Sciences, Exhibition Road, South Kensington, London, SW7 2DE, and from **Her Majesty's Stationery Office** at the addresses on Cover Page IV or through booksellers.

Institute of Geological Sciences
Exhibition Road
South Kensington, London, SW7 2DE

Geology of the Country around Okehampton

Royal 8vo, 256 pp., 20 text figs., 12 plates (8 colour and 4 monochrome).
Price £2·00 net

This book, the first memoir from south-west England since Geological Survey officers re-entered the region in 1962, was published in 1968 and deals with northern Dartmoor and its environs—a popular area of striking scenery and geological contrast. Geologist, geographer and naturalist—professional, amateur or student—will read it with advantage.

Detailed descriptions and interpretations of the nature and structure of the rocks are accompanied by discussion of the age, origin and emplacement of the Dartmoor Granite, of new palaeontological discoveries and their effect upon the classification of the Carboniferous rocks, and of geophysical studies and geomorphology.

Use of a smaller type size for 'details' within the text enables the general reader to reserve these sections for reference and yet gain a clear picture of earth history in the area, of the relationship of the rocks to land form and land use, and of natural resources in relation to past and future.

Free list of Institute of Geological Sciences publications is available from Her Majesty's Stationery Office, PM1C, Atlantic House, Holborn Viaduct, London, EC1P 1BN

HMSO

Government publications can be purchased from the Government Bookshops at the addresses on Cover Page IV (post orders to P.O. Box 569, London, SE1 9NH) or through booksellers.